Franklin Leonard Pope

Evolution of the electric incandescent lamp

Franklin Leonard Pope

Evolution of the electric incandescent lamp

ISBN/EAN: 9783337159481

Hergestellt in Europa, USA, Kanada, Australien, Japan

Cover: Foto ©berggeist007 / pixelio.de

Weitere Bücher finden Sie auf **www.hansebooks.com**

EVOLUTION

OF

THE ELECTRIC

INCANDESCENT LAMP

BY

FRANKLIN LEONARD POPE.

. The difficulties, apparently so insuperable, melted away. The electric lamp was completed. A piece of *charred paper* cut into a horse-shoe shape, so delicate that it looked like a fine wire, . . . proved to be the long sought combination. [*Edison's Electric Light*, by FRANCIS R. UPTON in *Scribner's Monthly*, February, 1880. Certified by Mr. Edison to be "correct and authoritative."] (p. 44).

. , I think it is clearly and fully shown that Sawyer and Man were the first inventors of the incandescent conductor for an electric lamp formed of *carbonized paper*. [Decision of E. M. MARBLE, Commissioner of Patents, in Interference between Sawyer & Man and Edison, 1883.] (p. 59).

ELIZABETH, N. J. :

HENRY COOK, PUBLISHER,

1889.

The outcome of a race of diligence between two independent but equally meritorious inventors, is perhaps as often as otherwise determined by chance or accident. In this respect, it may not inaptly be compared to the result of a horse-race, in which the fortunate winner carries off, not only all the honors, but the purse as well, although his nose may have passed under the wire barely an inch in advance of some of his no less deserving competitors. It is a matter of common observation that when the fullness of time arrives, the discovery or invention for which the world has been waiting is certain to be made. The critical student of affairs perceives, that however wonderful or however unexpected that invention may appear, it is seldom that it is not found to be a necessary sequence of a long series of other discoveries and inventions which have preceded it. Even in those rare cases in which an improvement of indisputable novelty and originality is made known to the industrial world, it is scarcely ever sufficiently perfected in its details to be capable of practical use until it has been worked upon and improved by many hands and many minds.

But it has always been the way of the world to consider every such invention, especially when of a character to appeal to the minds of the masses, or to identify itself closely with the everyday life of the community, as the work of some particular individual, who, as it were by common consent, is regarded as its sole originator and contriver, and upon him, fame, honor and wealth are lavished without stint, in childlike unconsciousness of the universal truth that inventions of this character are not made, but grow; that they are not the fruit of momentary inspiration, but on the contrary, are the inevitable results which from time to time mark the slow but constant progress of scientific and industrial evolution.

The history of electric illumination forms no exception to the general law, but the materials for that history are hidden in the voluminous records of the Courts, and of the Patent Office; in the files of newspapers, and in the transactions of learned societies.

In the following pages an effort has been made to bring together, in a convenient and accessible form, some of these detached fragments which seemed to have present or prospective historical value, and at the same time, as far as practicable, to indicate for the benefit of future investigators the sources from whence they have been derived. It is hoped that each reader will thus be enabled to formulate his own answer to the inquiry as to whom the world is principally indebted for modern incandescent electric lighting.

In the art of incandescent electric lighting, the lamp is the keystone of the entire structure. The one essential feature which differentiates the modern and successful lamp from its unsuccessful predecessors, is *the arch-shaped illuminant of carbonized organic material.* Before this invention was made, incandescent lighting was a commercial impracticability ; no sooner had it been made than the ultimate success of the scheme became not merely a probability but a certainty.

In the hands of the trained electrical engineer and the skilful mechanician, the complete and perfect result which we see to-day has been but the logical and necessary outcome of that fundamental discovery.

Elizabeth, New Jersey, U. S. A.,
    September 5, 1889.

CONTENTS.

CHRONOLOGY OF THE MODERN INCANDESCENT LAMP.

INCANDESCENT ELECTRIC LIGHTING.

INTRODUCTORY.

It is just half a century since the Congress was in session before which Morse and Vail made the first public exhibition of the electric telegraph; the earliest, and for a long series of years the only important application of electricity to the useful arts.

It is difficult to realize that but little more than a decade has elapsed since the illumination of the streets of Paris by the Jablochkoff electric candle, during the Exposition of 1878, gave the renewed impulse to the study of electricity, which has resulted in the development of a long and brilliant series of inventions, already destined to make this subtle form of energy one of the controlling factors of modern civilization.

In a recently published work on the economic value of electric light and power, the following statement occurs, the force and truth of which will be instantly recognized by every thoughtful mind :

So much has been done in applying electricity to the uses of light and power, that the following statement of some of its possibilities is no longer prophetic

Electric light and power will produce more changes in the mechanical servants and conveniences of civilized life than has ever been caused by the use of any other method or force which has been subjected to the service of man.

The uses of Electricity are limited only by the want of intelligence in producing fitting apparatus for its application. The demand for such apparatus is limited only by the want of intelligence on the part of the public to know how to use it properly.

So rapid has been the development of the electric industry, few have been able to keep trace of its achievements, and fewer still have sounded its possibilities for the purpose of giving direction to its growth

It is now known to be practicable to distribute from one central station all the light and mechanical power used in any city. Not only this, the light and power can be **delivered to any point in the city; and more important still, it can be divided,** and delivered in **any quantity** needed.

From the same central station, the same generator in the station, if need be, accompanying each other over the same conductor—as it were one spirit with two forms of expression—the electric current places at the command of the poorest member of society **two basic requirements** of civilized life,—**light** and **power.** They come without noise, without heat, smoke, dust or dirt of any kind. They consume no oxygen from the air; they bring no poisonous gases into rooms; they destroy no property; they make life more healthful, comfortable, and freer from liability to accident; they are brought into use without effort; with silent, patient energy they wait the will of their employer.

But turn a key, and light appears where the sun cannot shine.

But turn a key, and the tireless energy of the universe will play alike with the smallest toy or the heaviest machine.

Did civilization advance when a machine gave partial liberty to the slave who sang "Song of the Shirt?" How much more shall civilization advance when that same slave gains the greater liberty of having light with which to see to work, and of having the mechanical force required to operate that same machine come to her in an electric current through a small flexible wire?

Comply with its conditions, and then but turn a key, and the servant of all life will be present in light and power. [1]

In January, 1878, a prominent English technical journal published a *resume* of the progress of electrical science during the preceding twelve months. The following extract from this article will serve to show the condition, at that date, of electric lighting industries on the other side of the Atlantic.

The electric light has been conspicuously before us during the past year, and has made some decided advances both in the way of improvement and in its practical application. This activity has been almost exclusively confined to the Continent, where a great number of experimental trials have been made, and works, warehouses, promenades, railway stations, ships, and locomotives lighted by its means. In England it has been applied to transatlantic steamers, and to ironclads as a means of defense against the attacks of torpedo-boats made under cover of darkness. In June last, the new "electric candle," with the kaolin wick, of the Russian engineer, M. Jablochkoff, was publicly exhibited at the West India docks, and was considered to be a striking success. In France, the magneto-electric machine of Lontin, by which a number of separate currents of different strengths are generated and distributed to separate circuits in order to feed a separate light in each, has been successfully tried and tested in a long series of experiments made at the Lyons railway station in Paris, and we believe it has been finally adopted there.

In connection with lighthouse illumination by electricity, the most notable event of the year is the comparative trial of the magneto-electric machines of Siemens, Gramme, and Holmes, made at the South Foreland lighthouse for the Trinity House Board. These trials established the decided superiority of the Siemens' machine over those of Gramme and Holmes which were experimented upon. [2]

When the possibilities of illumination by electricity dawned upon the minds of men, their imaginations were so captivated that it was not easy at first to look beyond in

---

[1] *Economic Value of Light and Power.* A. R. Foote, Cincinnati, 1880, p. 8.

[2] *The Telegraphic Journal,* January 1, 1878.

order to form an adequate conception of the far greater poten-
tialities of *electric power*. Hence the history of the earlier
years of modern electrical development, comprises for the
most part little more than a history of electric lighting.

Prior to 1877, the dynamo-electric machine, as a source of
almost illimitable electric power, had been brought to great
perfection. It had already become a standard article of manufacture and sale ; in Europe by Siemens [1] and Gramme [2]
and in this country by Wallace [3] Weston [4] Hochhausen [5]
and Brush [3]. *The problem of the economic production of electricity by power in any required quantity was practically
solved* [6]. The remaining problem confronting the inventor
of 1877, who sought to render it possible to utilize the
potent energy of electricity for the varied requirements of
modern civilization, was three-fold, namely :

　　　1.　DISTRIBUTION.
　　　2.　UTILIZATION.
　　　3.　MEASUREMENT.

*To what person or persons ought the credit of solving this mo-
mentous problem to be awarded?*

In the following pages it is proposed to throw some new
light upon this vexed question, as reflected from contemporaneous records and publications.

THE EARLY WORK OF SAWYER AND MAN.

To the summary which has been quoted from the English
journal should be added a brief statement of what was being
done in this country during the same period.

At least as early as June, 1877, William Edward Sawyer,

---

[1] Report of the Secretary of the Navy, Washington, 1876, p. 159. *Expose des Applications
de L' Electricite,* v. 526 (illus). Th. Du Moncel. Paris, 1878.

[2] *Engineering,* Nov. 27, 1874. Art. " The Gramme Electro Magnetic Machine " (illlus.).

[3] *Report of Secretary of Navy,* Washington, 1876, pp. 158-9. *Johnson's Cyclopædia,* 1878 .
Art. Magneto Electricity in appendix, by Prof. G. F. Barker.

[4] *Scientific American,* Sept. 2, 1876. Art. " The Weston Dynamo Electric Machine " (illus.),
p. 150.

[5] Trade circular of William Hochausen, 1876. See Comp'ts Record, *Weston v. Arnoux &
Hochausen.* Evidence of W. B. Hollingshead, pp. 41, 50, 222.

[6] So far as the production of currents is concerned, I think that in the dynamo machines
of the present day we have reached almost perfection. Siemens' machine reproduces in the
form of electric current 90 per cent of the mechanical energy thrown into it ; and that is
as near perfection as any machine has ever obtained. Gramme's machine approaches that
very closely ; and therefore so far as the production of current is concerned there is nothing
wanting. [Evidence of *W, H. Preece,* Electrician to the Post Office. Parliamentary report on
Electric Lighting, May 2, 1879, p. 65].

a native of New Hampshire, who had been for some years a telegraphic operator in the New England states, and subsequently a reporter and journalist in Washington, D. C., directed his attention to the making of inventions in electric engineering and electric lighting. The first fruits of his labors which are matters of public record are found in his patent for an Improvement in Electric Engineering and Lighting Apparatus and System, for which his application was filed June 27, 1877, and granted August 14, 1877. The following verbatim extracts from the specification of this pioneer patent will show the direction of Mr. Sawyer's early work.

The object of my invention is to supply the streets, blocks or buildings of a town or city, in a practicable manner with any desired quantity of electricity, for the purposes of electrical illumination, electroplating, electric heating, the running of electro-magnetic engines, &c. I place the generator or generators of electricity in any convenient portion of a locality, whence I carry the necessary conductors over or under ground to the streets, blocks or buildings in which the electric current is to be utilized.   .   .   .   .

The advantages of my invention are that it enables householders to obtain a supply of electricity for any purposes without the care and inconvenien e attending the maintenance of local galvanic batteries; that it greatly reduces the cost of electricity to consumers, and, lastly, that it renders practicable the lighting of buildings by electricity.   .   .   .   .   .   .

The object of my invention is to supply streets and houses with electricity for local use or work, the same as gas or water is supplied to streets and houses at the present day, so that whenever a flow of electricity is desired it is only necessary to open a stop-cock, so to speak, and obtain whatever supply is needed.

A week later than this, August 21, 1877, Sawyer obtained a second patent for an Electric Lamp or Burner, which, although a crude and impracticable method of applying the electric current for illuminating purposes, is important, not only as showing the line of experiment and investigation which was being pursued by him, but because of other considerations which will appear further on.

The character of this invention will be understood from the extracts from the specification which follow :

My invention consists in an arrangement and combination of parts whereby I am enabled to place a considerable number of electric candles in a single circuit, and to dispense with the use of carbon points ordinarily employed in electric lights. I cause the electric current to heat to incandescence a platina wire or wires, by the bearing of which against preferably white refractory substances, such as clays, lime, &c., the heat is transferred to those substances, and a soft glowing light results. I do not, however, limit myself to the employment of platina wire, or to the heating of any particular substance.

The claims of this patent are :

1. The method of obtaining an electric light, consisting in heating a refractory substance by bringing it in contact with a heated conductor of electricity.

2. An electric candle, in which a conductor of electricity, heated to any desired degree of intensity, renders luminous a non-conducting substance in contact therewith, as set forth.

In a subsequent patent (No. 196,834) for which the application was filed August 10, 1877, and issued November 6, 1877, we find Sawyer still further developing his original conception of a general system of electrical distribution, as is evidenced by the following extracts from the specification :

Be it known that I, WILLIAM EDWARD SAWYER, of the city, county, and State of New York, have invented certain new and useful improvements in *Electric Engineering and Lighting Systems*, of which the following is a full, clear, and exact description.

My invention relates to electrical arrangements whereby a number of electric lamps or electric engines may be placed in the circuit of a single conductor, or operated from a generator common to all of them, with facility, in a practical manner, and without interference one with another.

In Letters Patent heretofore granted to me, I have described the general features of my electric engineering and lighting system.   .   .   .

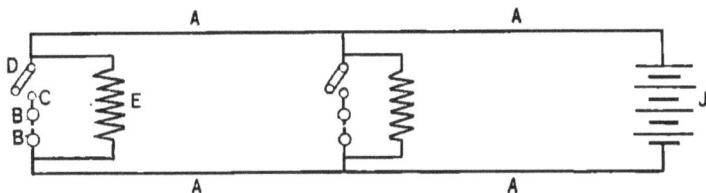

Fig. 3.

Diagram of parallel branch circuits, from Sawyer's patent.

In the system of branch circuits shown in Fig. 3, (the arrangement of which I do not claim as any part of my invention, excepting in combination with my circuit-continuity-preserving devices. whereby the current flowing through the branches is locally changed from a non-working to a working course without disturbance of the branch or main circuit, I have shown an application of the devices shown in Fig. 1.   .   .   .   .   .   .

It is apparent, further, that to insure the requisite conductivity for the great volume of current required, the main conductors must be very large, and as a single conductor traversing a territory Letters Patent No. 194,111) is forced to take many devious paths, passing into and out of houses, or into and out of rooms, to and from lamps or engines, in all of which deviations the large size of the conductor must be maintained, the expense of construction is very great. By placing my circuit-continuity-preserving devices, however, in branch circuits, the branches alone pursuing the devious paths mentioned in running across from the two large mains or conductors, the branch conductors may be small, for if I wish to divide the current flowing in the two main conductors, and thence through the branches among, say, fifty branches, the wire forming the

branches need be only one-fiftieth the mass of that forming the two main conductors; hence, instead of running a single very large and expensive conductor
into and from houses to and from each lamp or engine, I have only to run two
main conductors, and by small and inexpensive branches reach the lamps or engines.
The current is evenly distributed throughout all of the branches, and my circuit-
continuity-preserving devices, whether those of my present invention or those
of Letters Patent No. 104,111, or any others of like function, enable me to light
or put out any one or more of the lamps, or start or stop any one or more of the
engines operated by the current common to all, without decreasing or increasing the flow in the other branches.

It is important to note, *first*, that the object which the inventor designed to accomplish was to operate a large number
of lamps from a common generator without interfering with
each other ; in other words, the *division of the electric light*, and,
second, that in carrying out his invention, he describes what
he calls " a system of branch circuits," that is to say, *the identical arrangement of the lamps in parallel with the generator, now
in universal use*, at the same time admitting it to be old, and
stating that he does not claim it as part of his invention, except in connection with his own devices.

While engaged during the year 1877 in working out these
inventions, Sawyer's immediate resources were but limited,
and, as is frequently the case with inventors, he was much
hampered in his progress for want of means. In January,
1878, he chanced to make the acquaintance of Albon Man, a
middle-aged lawyer of Brooklyn, who had taken the natural
science course when in college, and who had never ceased to
retain a deep interest in scientific matters. Mr. Man became
greatly impressed with the possibilities of Sawyer's work, and
his interest in the matter resulted in a partnership, the original
design being that Mr. Man should furnish money to enable
Sawyer to complete his inventions. Early in February a shop
was accordingly fitted up at 43 Centre Street, New York. Mr.
Man almost immediately became personally interested in the
work of invention, and thereafter it was carried on jointly by
the two partners. As early as June, 1878, three patents were
issued to Sawyer and Man as joint inventors. One of these,
No. 205,144, was for an improvement in Incandescent Lamps,
in the specification of which occurs the statement :

At the present day *it is not new* to produce a light by causing the electric
current to heat a carbon conductor to incandescence in a vacuum.

No. 205,303, dated June 5, 1878, is a patent for an improvement in electric lighting systems, the following extracts from which will serve to show its general scope :

In another application filed by us, we have shown and described an electric lamp in which light is produced without consumption of the carbon producing it.

The object of our present invention is to provide for such a distribution and application of the electric current as will render our own or any other lamp practically operative upon an extensive scale.

Our system consists in (first) a new arrangement of electrical circuits, which are in part (second) composed of conductors **diminishing in transverse mass as they increase in length.**

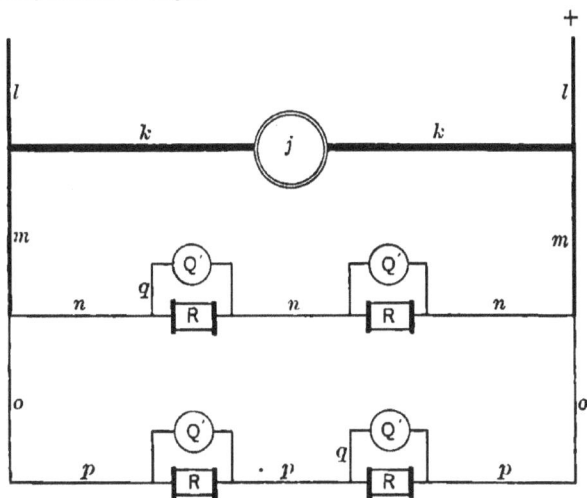

Fig. 5.

Conductors diminishing in transverse mass as they increase in length (from Sawyer & Mann's patent, 205,305) ; *j*, dynamos ; *k k*, feeders ; *m m*, *o o*, decreasing mains ; *n n n*, *p p p*, branches . Q Q Q Q, lamps in parallel series.

In the lamp of our invention hereinbefore mentioned, which we prefer in all cases to employ, a small piece of **carbon is heated to incandescence** in an atmosphere with which it will not chemically combine.

It will be observed that this patent clearly shows and describes a number of the essential features of the modern systems of electrical distribution for incandescent lighting, among which are—

First—An electric lamp in which *light is produced without consumption of the carbon.*

Second—The arrangement of *the lamps* or translating devices *in parallel arc with the generator ;* and

Third—The device of using *parallel main conductors* which *diminish in transverse mass* as they *increase in length.*

The substantial progress which had been made by Sawyer and Man in the development of a general system of electric light and power distribution, had by this time demonstrated that the scope of the invention was too broad to be successfully handled by individual enterprise. Mr. Man therefore sought to interest others in the undertaking, with the result that on July 8th, 1878, the "Electro-Dynamic Light Company of New York," was incorporated for carrying on the business. It is worth while to note the objects of this company as stated in its articles of incorporation, viz.:

The production of light and power by means of electricity; the lighting of streets, buildings and other places ; producing, conducting, and distributing electrical currents for lighting and other purposes, and the manufacture and sale of all machinery necessary for and adapted to accomplish the purposes named. [1]

In patent No. 205,305 of Sawyer and Man, dated June 25, 1878, we again find the lamps arranged in parallel with the generator, and n combination therewith another most important feature, which may be regarded as the very foundation corner-stone of modern methods of electrical distribution, and that is the *regulation of the production of electricity at its source,* so as to supply the *exact quantity* needed for the number of lamps alight at any time. It follows, as a matter of course, when the lamps are arranged in parallel, that if the proper quantity of electricity is supplied, it must distribute itself equally among the whole number of lamps and hence a *constant illumination* will be maintained in every part of the system. [2]

---

[1] The incorporators of the Electro-Dynamic Light Company were Albon Man, William E. Sawyer, Hugh McCullogh, Lawrence Myers, Jacob Hays, James P Kernochan and William H. Hays. Its nominal capital was $10,000, with $290,000 of scrip.

This was the *earliest* company incorporated for the purpose, among other things, of carrying out a *general system of incandescent electric lighting.* Thomas Wallace, of Ansonia, Conn., became the President of the company May 19, 1879. Owing to internal dissensions between the stockholders and trustees, which will be referred to elsewhere, the business did not prosper, and on April 6, 1881, the patents and property were sold to the Eastern Electric Manufacturing Company of Middletown, Conn.

THE CONSOLIDATED ELECTRIC LIGHT COMPANY, of New York, organized September 8, 1882, acquired the patents and property of the Eastern Electric Company, and in 1888 the control of the Electro-Dynamic Light Company also passed into the hands of The Consolidated Company, Mr. Wallace having previously resigned its presidency. The Consolidated Company has since passed under the control of the WESTINGHOUSE ELECTRIC COMPANY, of Pittsburgh, which has thus become the legitimate successor of the Electro-Dynamic Light Company of 1878.

[2] More than a year later than this, Sir William Thomson, in giving evidence on May 24, 1879, before a Select Committee of the House of Commons, said : "With proper regulators, by *inventions not yet made,* he machine will have a governor, according to which, when you short circuit without resistance any one of the lights, the machine will not give more current than you want for the other light or lights, whether in the same circuit or in parallel circuits" *Minutes of Evidence,* p. 187.

The importance of this improvement will be understood from the extracts from the specification which follow :

To all whom it may concern:

Be it known that we, WILLIAM EDWARD SAWYER, of the City, County and State of New York, and ALBON MAN, of the City of Brooklyn, County of Kings, and State of New York, have invented certain new and useful Improvements in Regulators for Electric Lights, of which the following is a full, clear and exact description:

The objects of our invention are, first, to supply to an electrical circuit or circuits containing electric lamps an **absolutely uniform volume of current,** in order to obviate the unsteadiness which has characterized most attempts at distribution; second, **to supply, automatically, only so much current as may be necessary to meet the demand,** thereby insuring the economical and satisfactory operation of a distributing system; and, third, when the demand for the current is lessened or increased by the removal of lamps from or their introduction into the circuit or circuits supplied by the generator, to cut off steam from, or add steam to, the engine driving the generator, **in order that there may be no waste of power** when the demand for the current calls for a limited application of it merely.   .   .   .   .

The regulator of our invention so controls the steam or other power actuating the generator of electricity that this power is increased or diminished, as the demand for the current is increased or diminished; and, while we have shown the result as accomplished in one way, we do not confine ourselves to the precise apparatus and the manner of its operation shown and described.

We prefer to place a sensitive magnetic apparatus in the circuit of the generator, which, when a lamp is lighted, so to speak, automatically takes cognizance thereof, and supplies that quantity of steam to the engine which is necessary to drive the generator at that speed which will give the exact required volume of electricity. We may so arrange our apparatus, however, that the regulator shall start different generators or couple different parts of a single generator in circuit, as may be required to supply the current; or we may arrange the regulator to control the current by the automatic changing of resistances. We may, in fact, in many ways, naturally suggestive, vary the application of our invention; and hence we do not limit ourselves to the employment of any special form of apparatus   .   .   .   .   .   .

When the circuits are parallel circuits, as shown, and the current is divided among them, it may be preferable to   .   .   .   .   increase the quantity of current supplied, by coupling together several generators or parts of a single generator, rather than to increase the speed of the engine.

When the lamps are arranged in series, so that the current passing into and out of one lamp enters another, and so on, the introduction of a lamp, inasmuch as it increases the resistance of the circuit, and thus renders an increase in the tension of the current necessary, should preferably have the effect of increasing the speed of the generator.

The principal claim of this patent is in the following words :

In an electric lighting system, a generator of electricity, two or more electric lamps supplied thereby, and an independent electro-magnetic switch or regulator, energized by the current from the generator, and acting automatically upon the occurrence of any change of electrical condition in the circuit or circuits supplied, to allow of a corresponding change in the quantity or intensity of the current generated.

It was not until after Sawyer, first by himself, and afterward in conjunction with Man, had thus far developed his original conception of *a general system for the electrical distribution of light and power from a central station*, that Edison appeared in the field.

THE EARLY EFFORTS OF EDISON.

The earliest record of the name of Thomas A. Edison in connection with electric lighting appears to have occurred in an article printed in the *New York Sun* of September 11, 1878, from which the following is extracted:

While visiting the mining regions of the Sierra Nevada and Rocky Mountains in his late western tour, Professor Edison was struck with the difficulty there had by miners in drilling and boring, though in many cases in the vicinity of rapid-flowing mountain streams. Except in "placer" mining, where the ore is washed out of the bed or banks of a river, or where expensive steam drills are used, the work of mining is laborious. While watching miners drilling by hand, a means of facilitating this work evolved itself from Edison's fertile brain. Turning to his intimate friend, Professor Barker, of the University of Pennsylvania, he exclaimed abstractedly: "Why cannot the power of yonder river (pointing to the Platte River on the plain a thousand feet below) be transmitted to these men by electricity?" This thought seemed not to go from Edison's head, and all the way across the plains on their journey home he and his friend "Barkey" discussed various problems for the transmission of power.

Before starting for the West Professor Barker had visited Ansonia, Conn., where his old friend, Mr. William Wallace, is engaged in the manufacture of electrical machines. Mr. Wallace has profound scientific research, and besides the mechanical part of his business, he devotes himself enthusiastically to that part of physics that comprehends electricity, magnetism and the polarization of light. He showed Professor Barker an instrument to which he had devoted the best years of his life, but which was yet in a crude condition. He is still experimenting with it, but he believed he would so perfect it as to transmit power from one point to another by means of electricity.

When the dison party had arrived in New York, Professor Barker bethought himself of the instrument previously shown him by his friend, and which at the time he had only cursorily examined. He invited Professor Edison to visit Ansonia with him, an invitation that was at once accepted. Last Sunday [September 8, 1878,] was the day fixed upon. The party consisted of Professors Edison and Barker, Professor Chandler, of the Board of Health, and Mr. Edison's assistant, Mr. Batchellor.

It was an agreeable surprise to the party to find that Mr. Wallace had perfected his machine. Being exceedingly modest, and caring not for notoriety, he had shown the instrument to few, and these only persons whose lack of scientific knowledge prevented them from comprehending its usefulness. Mr. Wallace calls it a telemachon, and he smiled with pride as he pointed to a number of these machines, each one an improvement upon its predecessor, and each having required years to design, and nearly $1,000 to construct.

Mr. Edison was enraptured. He fairly gloated over it. Then power was applied to the telemachon, and eight electric lights were kept ablaze at one time, each being equal to 4,000 candles, the subdivision of electric lights being a thing unknown to science. This filled up Mr. Edison's cup of joy. He ran from the instruments to the lights, and from the lights back to the instruments. He sprawled over a table with the simplicity of a child, and made all kinds of calculations. He calculated the power of the instrument, and of the lights, the probable loss of power in transmission, the amount of coal the instrument would save in a day, a week, a month, a year, and the result of such saving on manufacturing.

That a man like Mr. Wallace, after studying privately on the subject for years, should calmly, deliberately, and without ostentation, bring out before them an instrument calculated to revolutionize the entire manufacturing business, filled the party with amazement.

By means of it power may be obtained from places where river power or tidal power is abundant, or may be generated where fuel is cheap, as at the coal mines, and by means of an ordinary cable be transmitted hundreds of miles. The cable may be tapped at any point, and power used therefrom.

.   .   .   .

Mr. Edison believes that he can so assist Mr. Wallace in perfecting the telemachon that power may be transmitted from one point to another as though it were a telegraphic message. Already, by means of this instrument, Mr. Wallace is enabled to transmit the power of the Naugatuck river a quarter of a mile. The power of this stream is great enough to drive the ponderous machinery of the Wallace factory, where 300 men are employed. A series of experiments with the instruments has shown that *in the transmission of this enormous power by electricity only 20 per cent. is lost. The electricity from the wonderful telemachon may be applied to illumination.* It solves the problem of the subdivision of electric lights.

The following from the *Sun* of September 16, 1878, graphically relates the progress of invention at Menlo Park during the succeeding week :

Mr. Edison says that he has discovered how to make electricity a cheap and practical substitute for illuminating gas. Many scientific men have worked assiduously in that direction, but with little success. A powerful electric light was the result of these experiments, but *the problem of its division into many small lights was a puzzler.* Gramme, Siemens, Brush, Wallace, and others produced at most ten lights from a single machine, but a single one of them was found to be impracticable for lighting aught save large factories, mills, and workshops. It has been reserved for Mr. Edison to solve the difficult problem desired. This, he says, he has done within a few days. His experience with the telephone however has taught him to be cautious, and he is exerting himself to protect the new scientific marvel, which he says will make the use of gas for illuminating a thing of the past. While on a visit to William Wallace, the electrical machine manufacturer in Ansonia, Connecticut, he saw the lately perfected dynamo-electric machine for transmitting power by electricity. When power is applied to this machine it will not only reproduce it, but will turn it into light. Although said by Edison to be more powerful than any other machine of the kind known, it will divide the light of the electricity produced into but ten separate lights. These being equal in power to 4,000 candles, their impracticability for general purposes is apparent.   .   .   *Edison, on returning home after his visit to Ansonia, studied and experimented with electric lights.* **On Friday last** ⌜September 12, 1878] **his efforts were crowned with success,** and the project that has filled the minds of many scientific men for years was developed. "I have it now!" he said on Saturday, while vigorously turning the handle of a Ritchie inductive coil at his laboratory at Menlo Park, and, *singularly enough, I have obtained it through an* **entirely different process** than that from which scientific m n have ever sought to secure it. They have all been working in the same groove, and when it is known how I have accomplished my object, everybody will wonder why they have never thought of it, it is so simple. When ten lights have been produced by a single electric machine, it has been thought to be a great triumph of scientific skill. With the process I have just discovered I can produce 1,000—aye 10,000—from one machine. Indeed, the number may be said to be infinite. When the brilliancy and cheapness of the lights are made known to the public—which will be in a few weeks, or, just as soon as I can thoroughly protect the process—illumination by carburetted hydrogen gas will be discarded. With 15 or 20 of these dynamo-electric machines recently perfected by Mr. Wallace **I can light the entire lower part of New York City, using a 500 horse-power engine.** I purpose to establish one of these light centers in Nassau street, whence wires can be run up town as far as the Cooper Institute, down to the Battery and across to both rivers. These wires must be insulated, and laid in the ground in the same manner as gas pipes. I also propose to utilize the gas burners and chandeliers now in use. In each house I can place a **light meter,** whence these wires will pass through the house, tapping small metallic contrivances that may be placed over each burner. Then housekeepers may turn off their gas and send the meters back to the companies from whence they came. Whenever it is desirable to light a jet it will only be necessary to touch a little spring near it. No matches are required. "Again, the same wire that brings the light to you," Mr. Edison continued "will also bring power and heat. With the power you can run an elevator, a sewing machine, or any other mechanical contrivance that requires a motor, and by means of the heat you may cook your food. To utilize the heat it will only be necessary to have the ovens or stoves properly arranged for its reception. This can be done at trifling cost. A dynamo-electric machine called a telemachon, and which has already been described, may be run by water or steam power at a distance. When used in a large city the machine would of necessity be run by steam power. *I have computed the relative cost of the light, power, and heat generated by electricity* transmitted to the telemachon **to be but a fraction of the cost** where obtained in the ordinary way. By a battery or steam power it is **46 times cheaper,** and by water power probably **95 per cent. cheaper."** *It has been computed that by Edison's process the same amount of light that is given by* 1,000 *cubic feet of carburetted hydrogen gas now used in this way, and for which from $2.50 to $3 is paid, may be obtained for from* **12 to 15 cents.** Edison will soon give a public exhibition of his new invention.

An article of similar purport appeared in the *New York Tribune* of September 28, 1878, from which the following extracts are taken :

Thomas A. Edison is now engaged in making some new experiments with electric currents. A reporter of the *Tribune*, who visited the workshop at Menlo Park, N. J., found the inventor at work with a new apparatus which had been set up in the building on the previous day. It was a magneto-electric machine, made by Wm. Wallace of Ansonia, Conn. Mr. Edison recently visited Mr. Wallace's manufactory at Ansonia, and spent a day in examining the machines which have been constructed there. Mr. Wallace has manufactured electrical instruments for several years, and has spent much time and money in perfecting a magneto-electric machine which he calls the "telemachon." Its superiority over other machines of the same kind consists mainly in a nice adjustment of the several parts, and a proper proportion between the fields of force and resistance. It produces an enormous amount of electricity for the amount of force used in propelling it.

Mr. Wallace has made his machines so perfect that **nearly 80 per cent.** of the power used in propelling them is changed into electric-motive force. . .

Mr. Wallace's machines produce electricity which can be made available for **electric lights.** Mr. Edison continued. **" I have let the other inventors get the start of me in this matter somewhat, because I have not given much attention to electric lights; but I believe I can catch up to them now.** I have an idea that I can make the electric light available for all common uses, and supply it at a **trifling cost,** compared with that of gas. **There is no difficulty about dividing up the electric currents and using small quantities at different points.** The trouble is in **finding a candle that will give a pleasant light,** not too intense, which can be turned on or off as easily as gas. Such a candle can not be made from carbon points, which waste away and must be readjusted constantly while they do last. **Some composition must be discovered which will be luminous when charged with electricity, and that will not waste away.** A platinum wire gives a good light when a certain quantity of electricity is passed through it. If the current is made too strong, however, the wire will melt. I want to get something better. [1] *I have a chemist at work helping me to find the composition that will be made luminous by electricity.* Now that I have a machine to make the electricity, I can experiment as much as I please. *I think,"* he added, smiling, *" there is where I can beat the other inventors, as I have so many facilities here for trying experiments."*

---

1 In connection with this remark attributed to Edison, compare a report made two years earlier by the Electrician of the United States Torpedo Station at Newport, R. I., to the Hon. Secretary of the Navy, which contained the following passages in reference to incandescent electric lighting :

Incandescent Platinum. When either **platinum** or **iridium** are rendered **incandescent** by the passage of an electric current through bars or wires, made of either metal, a **mild and pleasant light is emitted,** much less contracted and glaring than the light obtained from carbon pencils ; with this advantage, no vitiation of the atmosphere occurs, and the **amount of light at any one point can be made as small as may be desired.**

Platinum affords about 100 candle-light per square inch of incandescent surface when within 220 degrees of the point of fusion, and a bar or wire of it **can be maintained at this temperature for any length of time** by means of **Farmer's automatic regulator,** controlling a suitable current,

Iridium, from its higher melting point yields more light per square inch of ignited surface, and can also be readily maintained at any desired temperature below fusion by means of the apparatus above mentioned. [The bars or strips actually used, are stated elsewhere in the report to have been about 2 inches long, 3-8 of an inch wide and 1-300 of an inch in thickness].
Report of the Secretary of the Navy, Washington, 1876, p. 161.

"If you can make the electric light supply the place of gas, you can easily make a great fortune," the reporter suggested.

"I don't care so much for a fortune," Mr. Edison replied, "as I do for getting ahead of the other fellows."

These statements do not indicate that Edison had at that time accomplished much of importance in electric lighting, but they do indicate that he was by no means ignorant of the work which Sawyer and Man had already done in the field which he proposed to enter, and if possible take possession of.

The latter part of the year 1878 was in Great Britain a period of unusual commercial depression. The feeling of gloom in business circles deepened day by day, culminating early in October, in the announcement of the failure of the Glasgow Bank for $30,000,000, which very nearly precipitated a disastrous financial panic. At this juncture the first public announcement that Edison had made a great discovery in electric lighting reached London, and produced an extraordinary sensation in commercial circles.

A special cable dispatch published in the *New York Herald* of October 11th, and dated the preceding day, says :

The London Stock Exchange continues in an exceedingly gloomy condition. There is a general feeling of apprehension, as though coming failures cast their shadows before. All stocks are drooping; especially is this true of British railway stocks, more particularly the L. & Y. roads, which are thought to be badly involved. In gas securities there is a perfect panic. Owing to the publication of Professor Edison's discovery for the distribution of the electric light, some gas stocks fell 12 per cent.

A London dispatch dated October 12, and published in the same paper on the 13th, says :

Commercial Gas shares are quoted 22½ lower; London, 15½; Gaslight and Coke, 12½; mperial Continental, 18½, and Metropolitan, 8.

A reporter was instantly dispatched to Menlo Park, then the headquarters of Mr. Edison, and the material obtained from him was published in the *Herald* of October 12, as follows :

The alarm among the English gas companies, and the panic in their stocks on account of Edison's last invention in electric lights, as narrated in yesterday's *Herald* cable dispatches, are not without good cause. *America's great inventor has, in truth, solved the problem which for years had puzzled the ablest electricians of the age.* He has successfully divided the electric light and has made it, for illuminating purposes, as far superior to gas as gas is to the tallow candle of the past. His marvelous invention gives a mild, soft, yet brilliant light, pleasing to the eye, clear, steady and without blemish, and at a cost less than one-third of that required for gas. Those who have seen the invention pronounce it wonderful. Already a company has been formed, composed of a number of wealthy capitalists, and Mr. Edison's lawyers are engaged in completing the final arrang_ments for the legal transfer. Before long, therefore, the work of introduction will begin. The patents for the United States have only just been granted, while those for the various countries of Europe have not yet been obtained. To-day the Professor's solicitors send by steamer the necessary documents to their London agents for procuring the British letters patent. The French and other patents will be applied for simultaneously with the English one. When word is cabled Mr. Edison that the patents have been granted he will throw his invention open to the public gaze; but until then he declines to make known its details, *his recent controversy with Professor Hughes over the alleged stealing of the microphone having made him cautious.* A *Herald* reporter learned yesterday the system proposed to be adopted after the patents are all granted. First, the Professor will light up all the houses in Menlo Park gratis, and from his laboratory watch the light's progress from night to night. When all is in readiness for general introduction, central stations will probably be established throughout New York City, each station controlling a territory of, perhaps, a radius of half a mile. Wires will then be run in iron pipes under ground, after the manner of gas pipes, connecting dwellings, stores, theatres and other places to be lighted. The gas fixtures at present used, ins.ead of being removed, will be utilized to encase the wire. In the place of the burner will be **the invention**, and meters will be used to register the quantity of electricity consumed. The form is not yet determined upon. The light is to be of the batwing, 15-candle power character. To kindle it a little spring is touched, and instantly the electricity does its work. The amount of light can be regulated in the same way as can that from gas. To turn off the light the spring is again touched, and instantly all is darkness. No matches being used, and there being no flame, all the dangers incidental to the use of gas are obviated. The light gives out no heat. It is simply a pure white light, made dim or bright at the fancy of the person using it. The writer last night saw the invention in operation in Mr. Edison's laboratory. The illumination was such as would come from a brilliant gas jet surrounded with ground glass, only that the light was clearer and more brilliant. 'When do you expect to have the invention completed, Mr. Edison?' asked the reporter. 'The substance of it is all right now,' he answered, putting the apparatus away and turning on the gas. 'But there are the usual little details that must be attended to before it goes to the people. For instance, *we have got to devise some arrangement for registering, a sort of meter*, and again there are several different forms that we are experimenting on now in order to select the best.' 'Are the lights to be all of the same degree of brilliancy?' asked the reporter. 'All the same.' 'Have you run across any serious difficulties in it as yet?' 'Well, no,' replied the inventor, 'and that's what worries me, for in the telephone I found about a thousand obstacles, and so in the quadruplex. I worked on both over two years before I overcame them.'

.   .   .   .

'Of all things that we have discovered this is about the simplest,' continued Mr. Edison, 'and the public will say so when it is explained. We have got it pretty well advanced now, but there are some few improvements I have in my mind.'

.   .   .   .

Leaving Mr. Edison to continue his work, the reporter accepted the invitation of Mr. Griffin, his private secretary, to view the generating machine.

'Mr. Edison has just purchased a new one,' he said, 'which gives much power, and in a few days we shall have a 50-horse power engine to work it, as we do not get enough power with the present one. As regards the generators,' continued Mr. Griffin, 'Mr. Edison has as yet given but little attention to them. The cost can, by an improved generat·r, be reduced to a much smaller figure.'

The publication of the foregoing article, with others of like character, in the leading New York newspapers, created an unexampled excitement in financial circles, and led almost immediately to the organization of the Edison Electric Light Company, which numbered among its incorporators prominent capitalists and lawyers, some of whom had long been identified with successful and profitable electrical enterprises.[1] It is said that the sum of $100,000 was at once placed at the disposal of the inventor by the company, to be expended at his discretion in the prosecution of his researches and the development of his enterprise.[2] It would be unprofitable to attempt to quote from the innumerable articles, interviews and paragraphs in reference to electric lighting set afloat in the newspapers during the latter part of 1878 and the early part of 1879. The most absurd and exaggerated statements were eagerly printed by the newspapers and still more eagerly swallowed by an insatiable public.[3] An examination of the records of the Patent Office, which are accessible to all, enables the actual line of research pursued at Menlo Park dur-

---

[1] The *Edison Electric Light Company*, capital $300,000, filed articles of incorporation in New York, October 17, 1878, the incorporators being Thomas A. Edison, Tracy R. Edson, James H. Banker, Norvin Green, Robert L. Cutting, Jr. Grosvenor P. Lowrey, Robert M. Gallaway, Egisto B. Fabbri, George R. Kent, George W. Soren, Charles F. Stone, Nathan G. Miller and George S. Hamlin. The objects of the company, as stated in the articles of incorporation, were "to own, manufacture, operate and license the use of various apparatus used in producing light, heat, or power by electricity." Several of the incorporators were well acquainted with Edison, and possessed unlimited confidence in his ability to solve the most abtruse and difficult problems in electro-mechanics by the sheer force of inspirational genius. Thus Edison entered the field with an equipment of practically unlimited resources, and with the most unexceptionable endorsers.

[2] *Scribner's Monthly*, February, 1880, p. 534. (Art. by Francis R. Upton.)

[3] Not all the journals of the day could be persuaded to take alleged American inventions seriously, as the following amusing skit, translated from *Le Figaro* of Paris, is evidence :—

The Bunkumphone.—The bunkumphone, of American origin, is an ingenious combination of the telephone, phonograph and other *phono-grips* designed for the exclusive use of financiers. It consists essentially of a metallic disc in a state of excessive vibration, and having the property of putting in movement many other discs of the same sort in our continent of simpletons, and of attracting these European discs by an action as energetic as it is well dissembled. This disc is actuated by submarine cables, the chief of which connects the city of New York with the Institute of France. The attracting disc is of silver, and it has the weight and external appearance of a dollar. The attracted discs resemble ordinary hundred-sou pieces

This apparatus is being experimented upon at this moment with great success. We have been shown at the *Figaro* office this same bunkumphone, as a marvel as inexplicable as the mystic cabinet of the Davenport Brothers We have dismounted the little machine, and after having seen what was inside, we have said :—"Gogo, friend Gogo, the Frenchman is a born fool, everybody knows that ; but he has not quite got used to American humbugs. Gogo, beware !"

ing this period to be definitely traced so as to determine with sufficient accuracy for the present purpose the actual results which were obtained.

The specification of his first patent was signed by Edison on October 5, 1878.[1] It describes nothing more than a thermostatic regulator intended to be attached to each individual lamp to prevent the overheating and fusion of its incandescent platinum burner. A second application, signed November 14, describes a device designed to accomplish the same result in a different m inner,[2] and this was followed by a third of like character, dated Dec. 3, 1878.[3]

So far as electric lighting by means of a platinum or iridium incandescent strip is concerned, it is hardly necessary to say that at the date ref-rred to it possessed absolutely no novelty. To go no further back, it had been used in the naval experiments at the Torpedo Station in Newport, in 1876, (see note, page 15) all its properties had been thoroughly investigated and were perfectly familiar to scientific men.[4] The

(see note, page 15)

[1] U. S. Patent No. 214,636, April 22, 1879.
[2] U. S. Patent No. 214,637, April 22, 1879.
[3] U. S. Patent, No. 218,866, August 26, 1879.
[4] In 1847 Professor J. W. Draper, of New York, made a very interesting series of investiga-tions on the heat and light evolved by platinum wires when traversed by powerful electric currents, and suggested that the currents might be regulated automatically. [Vide *Silliman's Journal*, (2d series) iv., 1847].
Perhaps the earliest mention of the incandescent electric light proper is in a paper by William Robert Grove, the inventor of the well-known Grove battery, in the *Philosophical Maga-zine*, third series, vol xxvii., p. 442, from which the following is an extract:—

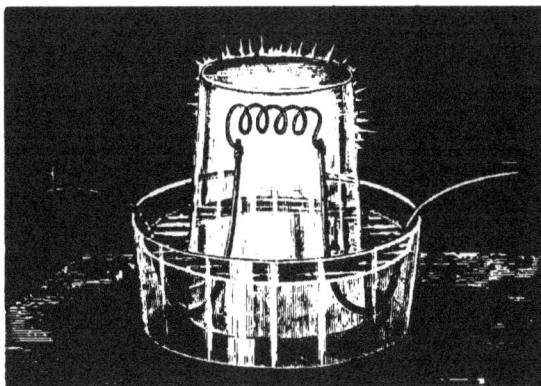

Grove's Incandescent Lamp of 1840.

The following was one of the apparatus which I used for this purpose, and by the light of which I have experimented and read for hours: A coil of platinum wire is attached to two copper wires, the lower parts of which, or those most distant from the platinum, are well var-nished ; these are fixed erect in a glass of distilled water, and another cylindrical glass over thermostatic regulators devised by Edison and described in his early applications for patents, in point of practical utility, were not for a moment to be compared with some which had preceded them—for example, that of Farmer, used at the Torpedo Station in 1875, and earlier,[1] and that devised by Sawyer and Man, which had already been patented and published, and which as previously stated, (page 10) operated in such a manner as to regulate the supply of energy at its source in accordance with the actual consumption, which is *now well recognized as the only feasible plan of electrical distribution* for incandescent lighting and power.

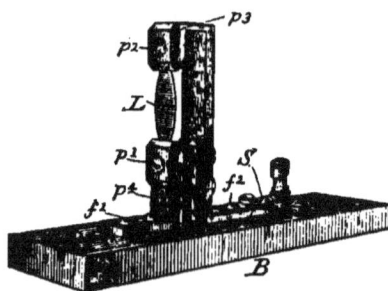

Farmer's Incandescent Lamp of 1859.

Even in the details of his apparatus for regulating the current traversing a platinum lamp, so as to prevent the melting of the burner, Edison was subsequently found to have been anticipated by Hiram S. Maxim, one of the early associates of Sawyer, (since become widely known as the inventor of the

them, so that its open mouth rests on the bottom of the former glass ; the projecting ends of the copper wires are connected with a voltaic battery (two or three pairs of the nitric acid combination), and the ignited wire now gives a steady light, or which continues without any alteration or inconvenience as long as the battery continues constant, the length of time being, of course, dependent upon the quantity of the electrolyte in the battery cells.

According to the statement made in the paper, these experiments were made in 1840 and 1841, shortly after the invention of the nitric acid battery. The publication above quoted was made in 1845.

[1] Under date of October 30, 1878, Professor Moses G. Farmer, of the U. S. Naval Torpedo Station, Newport, R. I., writing to the *Salem* (Mass.) *Observer* of November 2, 1878, says :

Some few of the citizens of Salem (among them ex-Mayor Williams, Mr. George D. Phippen Mr. J. H. Phippen, and perhaps others), will doubtless recollect a parlor at No 11 Pearl street, Salem, Mass , which was **lighted every evening during the month of July, 1859,** by the electric light, and this electric light was **subdivided** too ! This was nineteen years ago, and it was undoubtedly the first private dwelling ever lighted by electricity. . . . In the year 1875 I **subdivided an electric current into forty-two different branches,** putting a light into each branch. All these lamps were supplied with electricity from one machine, which did not weigh more than eight hundred pounds, and which was driven by a small steam engine.

A gentleman who saw this apparatus in operation has stated that the automatic regulator then used was so sensitive that the cooling of the platinum burners caused by the current of air in the room upon the opening of a door was sufficient to make it regulate the lights.

Maxim magazine gun), who actually constructed a platinum electric lamp with a thermostatic regulator in 1877.[1]

A significant comment·ry upon the character of the work which was being personally done by Edison in electric lighting at this time, is aff·rded by the specification of an electric generator, for which an application for a patent was signed by him on December 3, 1878,[2] which it is difficult for one possessing even a moderate degree of scientific knowledge to read without amazement. Setting out with the assumption that a vibrating body, such as a tuning fork or reed, can be maintained in vibration "by the exercise of but little power," he avers that by combining this device with a magneto-electric apparatus he is able "to obtain a powerful electric current by the expenditure of a small mechanical force."

This is perhaps the invention which is alluded to in an interview published in the *New York Sun* of Nov. 25, 1878 :

In further conversation, Mr. Edison said he was getting four times as much light with the same force as he did when he first began his experiments.   He found the generating machines in use faulty, and was making *two improved machines of his own*, with the view of turning the greatest amount of horse-power into electricity with the least possible loss.   These machines are to be especially applicable to his electric light.

The effect which this revelation produced upon the minds of the more intelligent of Mr. Edison's critics, is indicated in the following extract from the *Saturday Review* of January 10, 1880 :

The second patent then taken out by him was for a wonderful **dynamo-électric machine** of a wholly new construction.   We willingly give Mr. Edison credit for originality in this machine.   Coils were fixed to the vibrating arms of a monstrous tuning-fork, more than a yard long, and these, by the vibrations of the fork, were made to approach or recede from magnets, and th is currents were generated.   *If it were not actually in a patent taken out on Mr. Edison's behalf, all instructed persons would hesitate to believe that such an absurd arrangement could be seriously proposed*, at a time when such machines as the

---

[1] Mr. Maxim filed an application for a patent on December 23, 1879, in which he claimed "the combination of a thermostatic circuit regulator with an electric light having a continuous incandescent conductor," this claim being substantially identical with that of patent No. 214,636, which had been granted on April 22, 1879, to Edison.   In accordance with the statutes and the practice of the Patent Office, a proceeding, technically termed an interference, was instituted between Maxim and Edison to determine which was the prior inventor of the apparatus in controversy.   Testimony was taken by both parties, which proved that Maxim conceived the invention in November, 1877, and made drawings of it, and had an operative lamp constructed, which he successfully lighted with a battery in December, 1877, and with a dynamo machine early in the spring of 1878.   It was established beyond controversy that Maxim had fully completed the invention before Edison ever thought of it.   The proceedings terminated a decision on July 28, 1881, by acting Commissioner of Patents Stockbridge, awarding prior to Maxim, to whom a patent was accordingly issued on September 30, 1881. (No. 247,380.)

[2] U. S. Patent No. 218,166, August 5, 1879.

Gramme, the Siemens, the Lontin, the Brush, and a host of others were in existence, *much less that it could be proposed by a man of Mr. Edison's advantages and fame.* It is difficult adequately to express the ludicrous inefficiency of the arrangement; but one thing is abundantly certain, and that is that *the person who seriously proposed it was wholly destitute of a scientific knowledge of either electricity or the science of energy.* It is clear that he was tempted by the hope of getting out of the vibrations of the tuning-forks something more than the force he expended on them. No doubt he thought that vibration was so confirmed a habit with tuning-forks that they would vibrate on the merest hint being given to them. To those who remember the amusement that this wonderful invention excited among English electricians, it would be interesting to read the following passage from the latest authentic American account:

> Mr. Edison's first experiment in machines for generating the electric current did not meet with success. His primal apparatus was in the form of a large tuning-fork, constructed in such a way that its ends vibrated with great rapidity before the poles of a large magnet. These vibrations could be produced with comparatively little power. *Several works of practice* proved, however, that the machine was not practicable, and it was laid aside.[1]

We should very much like to know when these weeks of practice (not a very long trial for a new invention) took place. Not before the patenting, or it never would have been patented. Then it must have been after the patent was taken out—a matter which confirms the opinion held by most persons in England who were competent to judge of it, that no such machine had at the time ever been made (except, perhaps, on a small scale), and that the whole matter was a pure speculative suggestion Remembering the unrivaled opportunities for experiment possessed by Mr. Edison, the fact that he took out this patent without any adequate preliminary trial—and we are convinced that a most superficial investigation would have demonstrated its worthlessness—is a striking lesson as to the reliance that must be placed on the accounts of the extent of the preliminary experiments to which his so-called inventions are subjected.

If, therefore, the actual results which had been attained by Edison at the close of the year 1878 be summarized, it must be conceded that they comprised nothing more than the old idea of an incandescent wire of platinum or iridium, a useless thermal regulator designed to be attached to each individual lamp, and the preposterous tuning-fork dynamo. He was, as he had been from the very beginning, on a wrong track altogether. The triumphs of the future, as we now know, were destined to be accomplished by incandescent *carbon*, not by incandescent platinum, and by regulating the current at the *source of supply*, and not at the individual burners. In both these points Sawyer and Man were far in advance of Edison, for they had not only shown, in their patents already issued and published, an intelligent conception of the nature and scope of the problem upon which they were at work, but

---

[1] *New York Herald, Dec. 21, 1879.*

an equally clear prevision of the true path to be followed in the development of the incandescent electric lamp.

No better evidence of the truth of this assertion can be desired than is found in the following extracts from contemporaneous journals, comparing the respective opinions of Messrs. Sawyer and Edison at the close of 1878, upon the most available material for incandescent lamps:

SAWYER.	EDISON.
. . . . A year before Mr. Edison thought of it, I had a lamp using platinum, and operating on this principle, but *the cost of heating platinum is so much greater than the cost of heating carbon,* that I cast it aside as **utterly worthless.** The only thing we could liken it to was a hot poker.—[Letter of W. E. SAWYER in *New York Commercial Advertiser,* Nov. 23, 1878.] . . . . The very platinum lamp that was so much noised about, had been made a year before Edison's announcement of it, shown to everybody and cast aside as worthless by Mr. H. S. Maxim, of this city.—[See note p. 20, *ante*]. It was cast aside because platinum shortly disintegrates, and because it takes three horse-power to produce the lights of a first-class gas-jet from it with the best machines known. [Letter of W. E. SAWYER in *N. Y. World,* Dec. 23, 1879.	. . . **I use no carbon.** . . . . . *The carbon lamp won't do.* I have made repeated experiments with carbon.—[T. A. EDISON, interview reported in *New York Sun,* Nov. 25, 1878.] . . . It has been acknowledged by nearly all electricians that lighting by incandescence, especially the *incandescence of a metallic wire,* offers less obstacle to the division of the electric light than by any other method, and Edison believes it to be **the only reliable method,** because the light-giving metal is *an electrical constant,* whose resistance can always be *known and depended upon,* a condition which is exceedingly essential when many lamps must be supplied from one conductor.— *New York Herald,* Dec. 11, 1878. [Authorized article.]

Edison pins his faith to platinum, believes it to be "the only reliable method" and condemns carbon. Sawyer, having long since tested and abandoned platinum, gives his reasons therefor, and persistently continues his experimentation with carbon.[1]

---

[1] In confirmation of the correctness of Sawyer's opinion on this point, may be cited the views of the late Alfred Niaudet, a French physicist of deservedly high repute, who in a communication published early in 1879 in the *Cronica Cientifica,* expressed a very decided conviction that **carbon,** and not *platinum* or any other metal, is the best material for the production of the electric light by means of incandescence. In support of this position M. Niaudet adduces the following arguments: 1. At equal temperatures, carbon possesses a power of radiation in excess of that of platinum. 2. The heat capacity of platinum is far greater than that of carbon, and the former, consequently, requires the production of a far greater amount of heat than the latter to bring it to the required temperature. 3. Carbon is a worse conductor of electricity than platinum, and, consequently, may be of greater diameter without too greatly reducing the temperature. 4. Carbon is absolutely infusible, and may be brought to a white heat without interrupting the circuit by fusion.—[Vide *Tel. Jour.* Vol. vii, p. 119, Apr. 1, 1879.

Subsequent events have clearly shown that Sawyer and Man chose the right road at the beginning and never departed from it, while Edison, taking up the problem at a later day, spent more than a year in fruitless experiments in the wrong direction, the results of which were absolutely worthless, and at length had to be definitely and completely abandoned. At the end of 1878, as we now know, there existed no ground whatever for the repeated announcements in behalf of Edison that he had made a substantial advance towards the solution of the problem of general electric illumination, much less for the assertions, in many cases made by persons regarded as competent judges, that he had completely solved it.[1]

In a lecture describing Edison's inventions, delivered November 23, 1878, at Steinway Hall, New York City, Edward H. Johnson, since President of the Edison Electric Light Company, said:

Mr. Edison is a man of wonderful resources. When he starts to experiment he never follows in any rut made by anybody else, but always starts out on an entirely different line. I this week saw in his laboratory the mechanism for generating this current of electricity different from that used by anybody else.[2] Mr. Edison is subdividing the electric light, increasing the illumination and improving the mechanism for generating the current of electricity. He expects to be able to place 1,000 lights in the circuit. He knows the subject thoroughly and is on the track of every improvement in the way. He will not have as much experimenting with this as he had with the telephone. I feel justified in saying that within less than one year from to-night the electric light system of Mr. Edison will be in practical operation, not only in the streets and public houses alone, but in the private houses in New York city.

Early in the summer of 1879, Edison appears to have become alarmed at the possibility that the existing stock of platinum in the world would be wholly insufficient for the proper development of his invention. In this dilemma he felt that either a substitute for platinum must be found or new platinum

[1] As an example of the statements referred to, may be cited a lecture by Professor George F. Barker, of the University of Pennsylvania, delivered in November, 1878, in the course of which he said: "I hoped to be able to exhibit the famous light, but I am informed by Professor Edison that advices from his London solicitors prevent him making his invention public for twenty days yet, and I therefore have to wait; but within a week I have visited Menlo Park, and *after a thorough examination of Mr. Edison's discovery* I can say that **the problem has been solved**, and that Mr. Edison can place on every gas bracket and on every chandelier, burners which will give a brilliant white light, safe, pleasant, beautiful, and at about *one-third the cost now charged for gas.* **The practicability of the scheme is beyond question.**"—Vide *N. Y. Herald,* Nov. 22, 1878.

[2] If the lecturer referred to the tuning-fork dynamo mentioned on p. 21, *ante,*—of which the London *Engineer* said "Mr. Edison has probably succeeded in producing the very worst magneto-electric machine ever made,"—probably no one at the present day will feel justified in disputing the assertion.

mines must be discovered and developed. He determined upon the latter course, and at once dispatched agents to all parts of the world in search of the precious metal.[1]

It was an exceptionally fortunate circumstance for Edison that he was able about this time to secure the services of Francis R. Upton, a young man who before the completion of his college course, had exhibited mathematical ability of a high order. After his graduation, Mr. Upton spent two years in the physical laboratory of Professor C. F. Brackett, of Princeton, and another year in Berlin, under the instruction of the celebrated physicist Helmholz. He went into the employ of Edison, November 15, 1878, at which time the latter was still experimenting with the platinum spiral lamp and its thermal regulator.[2]

"Early in 1879," says Upton, "Edison determined the prerequisite of high resistance for a successful incandescent lamp, and his experiments during the year 1879, were largely directed to that end."[3]

The earliest recognition of the importance of high resistance in the lamp which appears in Edison's patents is contained in No. 227,229, for which his application was filed April 21, 1879, but which was not issued until May 4, 1880. This patent is the first which appears to have contained suggestions or combinations which ultimately proved of value in electric lighting. The following extracts from the specification will serve to indicate the progress which had been made by him up to this time.

---

[1] Recent reports from America state that the one obstacle met with by Mr. Edison at present to the complete success of his electric light is the high price of platinum for his incandescent coil. No other metal will answer, it seems, so Edison has sent off a dozen explorers to California, New Mexico and South America to ransack the gulches, with orders to find a large and constant supply of platinum. If they cannot find it on the surface of the earth, they have orders to proceed straightway to the interior.—*Tel. Jour.* Vol. vii, p. 267, Aug. 15, 1879.

[2] See evidence of Mr. Upton in Interference Record, *Sawyer & Man v. Edison.* Answer to Q. 2, 3, 5.

[3] This principle had, however, been familiarly known at a much earlier date. Its bearing upon the conditions of electric lighting had been pointed out by Professors Edwin J. Houston and Elihu Thomson, in a paper read before the American Philosophical Society Nov. 1, 1878, in which the following passage occurs:

In respect to the relations that should exist between the external and internal work of dynamo-electric machines, it will be found that the greatest efficiency will, of course, exist where the external work is much greater than the internal work, and this will be proportionately greater as the external resistance is greater.

The English patent of Lane Fox, No. 3988 of 1878, enlarges upon the same idea, as follows :

In order that the electric force may be conveyed at a high tension, that is, having a high electro-motive force, so that there may not be very great loss from the resistance of the conducting mains or conductors, I make the lamps when I use an alloy of platinum and iridium of lengths of fine wires, so that I may get a high resistance without having a large extent of luminous surface.

When platina and other metals that fuse at a high temperature are exposed to high heat and then cooled in the atmosphere, they are injured, so that they are not well adapted to use in electric lights for a long period of time.

I inclose the conductor that forms the electric candle in a transparent case and heat the same gradually to expel any gases from the material of the candle. I form a vacuum in the transparent case and then seal the same hermetically, so that all injurious atmospheric influences are avoided.

The invention further consists of a vacuum-receptacle made entirely of glass and sealed by melting the same in combination with an incandescent continuous conductor pyro-insulated.

The invention further consists in winding pyro-insulated wire upon a bottom of a compressed infusible substance, such as lime.   .   .   .   .

The drawing shows a section of the apparatus, in which B is the transparent bulb. This bulb is open at the smaller end and the burner *d* inserted, and the open end of the tube is placed in connection with a mercury vacuum-pump, the platina wires *g* and *f* passing through. The burner is connected with a battery and variable resistance coil while the vacuum is being made. The heat of the bobbin *d* is, in the course of one hour, brought gradually from the temperature of the air to vivid incandescence. When the vacuum is considered practically perfect the open end of the tube is melted and sealed. The platina wires passing through the glass are also sealed.

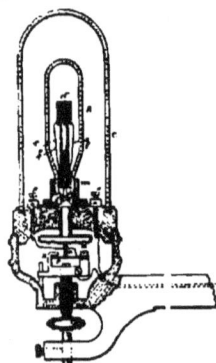

Thus I am enabled to obtain a nearly perfect vacuum, which is permanent, and at the same time give the platinum wire a new and unknown property of great value in electric lighting, which is, that a platina wire which melts in the open air at a point where it emits a light equal to four candles will, when operated upon as described, emit a light equal to twenty-five candles without fusion.

e is a cylinder, of lime, with a small spool on its extremity, on which the wire is coiled. About 30 feet of platinum or iridium wire coated with magnesia oxide is coiled upon the spool.

The wire may be of any size ; but I prefer to use wire .005 of an inch in diameter, which will give a resistance when incandescent of about 750 ohms. *By the use of such high resistant lamps I am enabled to place a great number in multiple arc without bringing the total resistance of all the lamps to such a low point as to require a large main conductor ;* but, on the contrary, I am enabled to use a main conductor of very moderate dimensions.

Another important point is gained by the use of lamps of high resistance, as the resistance of the wires leading from the main conductors may be of very moderate dimensions ; hence can be placed in the pipes already used for gas, and at the same time effect a great saving in the cost of wire.

Still another point gained is, that the high resistance of the lamps allows all to be placed in multiple arc, which is the only method where the maximum economy is attainable, as the lamps, when connected to the circuit, draw from the central station just sufficient current to maintain it at the proper tempera-

ture, and if by accident or want of regularity in the main current the strength
of the current should increase abnormally, the excess of heat sets the thermal
regulator in motion and disconnects the lamp entirely from the circuit, thus
stopping all further consumption of energy until the temperature of the lamp
is reduced to its normal conditions. . . . .

m is a lime cup, into which the small end of the burner is held. The
platina wires pass under it to the binding-posts H K.

In my improvement the chamber containing the light is made entirely of
glass, and I am able to obtain and maintain a vacuum, because there is no substance joined to the glass; hence the entire chamber can be hermetically sealed,
and the conductors of metal passing through the glass and around which the
glass is melted are so small as not to injure the glass by their expansion.

I am aware that carbon has been heated in the presence of both liquid and
gaseous materials for changing its character and adapting it to an electric
light. In my present invention the gaseous materials contained in metallic
wire are driven off by the action of heat evolved by an electric current while
the wire is in a vacuum, so that the pores of the metal are not filled with any
extraneous substance; but, on the contrary, the metal is solidified by the
removal of extraneous matter and the pores closed. . . . . .

I claim as my invention—

1. In an electric lamp, the combination, with a hermetically-sealed vacuum
chamber made entirely of glass, of metallic conductors passing through the
glass and around which the glass is melted, and an incandescent conductor
placed in the electric circuit, substantially as set forth.

It will be observed that in the lamp described in Edison's
patent, the illumination is designed to proceed principally
from a cylinder of lime, around and in contact with which
the incandescent conductor is placed. In this idea, he had,
as we have already seen, been substantially anticipated by
Sawyer.[1]

It has been claimed in behalf of Edison that the subject
matter of this patent, viz:—" an electric lamp having a continuous conductor (without regard to its material, resistance or
mode of preparation) and an exhausted glass enclosing globe[2] "
was original with him; but there would seem to be little invention displayed in the combination, in view of apparatus
which had already been employed by other physicists.

In 1877 and 1878, William Crookes, of the Royal Society,
published accounts of a remarkable series of experiments
on radiant energy.[3] Among the illustrative apparatus which
he constructed and employed was a little instrument which he

---

[1] Compare Sawyer's patent of August 21, 1877, p. 6, 7, *ante*.

[2] Circular of Edison Electric Light Company. July 27, 1882.

[3] See *Philosophical Transactions* of the Royal Society. clxiv. (1874) p. 501; clxv. (1875) p.
519; clxvi. (1876) p. 3.6; clxix. (1878) p. 244.

termed the radiometer, which may be briefly described as a small revolving wheel provided with vanes or paddles of mica, poised upon a delicate pivot within a highly exhausted and hermetically sealed glass bulb. Mr. Crookes showed that a paddle-wheel so constructed could be maintained in continuous revolution by the mechanical force of light alone. In an account of some of his experiments, published in London *Engineering*, April 19th, 1878, a description and illustration of one of these radiometers is given as follows :—

One of the instruments employed is that shown in Fig. 6, in which is suspended in the ordinary way upon a needle point a four-armed radiometer, carrying rectangular vanes of clear mica, b b, set on an angle with the horizontal arms like the vanes of a smoke-jack. A short distance below this fly is fixed in a horizontal plane a circular loop of platinum wire, c c, the two ends of which pass through the bulb, *being hermetically sealed into the glass*, and finish in the terminals $+$ and $-$, by which the platinum ring can be placed in circuit with a voltaic battery. *and thus rendered incandescent.*

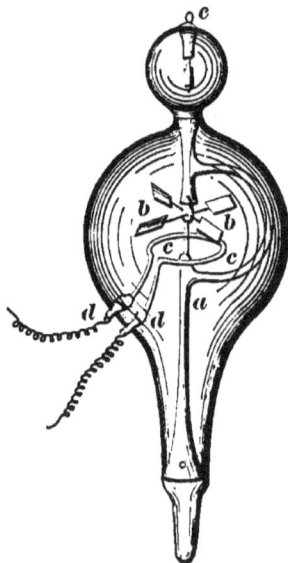

FIGURE 6.

For determining the pressure, the whole apparatus was connected by glass tubing and *fused joints* to a very perfect Sprengel mercury pump, designed and constructed by Mr. C. H. Gimingham, fitted with apparatus for obtaining a chemical vacuum by different absorptions of different substances contained in the residue gases, as well as for eliminating from the same all traces of aqueous vapor, and having attached to it McLeod and barometric gauges for measuring the pressure at different degrees of exhaustion. For the observation of the appearance of the electric discharge—a very valuable indication of the progress of the exhaustion—there was attached to the apparatus a vacuum tube in gaseous connection with its interior, and in which two platinum electrodes were placed very close together. In addition to these accessories, there was a small torsion balance carrying at one end a standard disc of blackened mica, and at its centre a small mirror, by which its angular displacement could be measured by the movement of a spot of light along a scale, and lastly, there was the special apparatus which has been described, containing the rotating disc and vanes, and the platinum ring which, by being put into and out of circuit by a Morse key with a voltaic battery through various resistances, could be rendered either just warm, or heated through various degrees of incandescence almost up to its temperature of fusion ; and when it is remembered that provision had to be made for the introduction of various gases into the apparatus previous to the process of exhaustion, it will be readily understood that, although the whole arrangement was as simple as its requirements would admit, still it was considerably elaborate ; from its necessary complexity, it required the utmost skill to construct, and it was impossible to be carried about.

In order to show the phenomena which have been described to the members of the Royal Society, in illustration of his paper, Mr. Crookes had several instruments identical in construction to that represented in figure 6, but containing gaseous vacua of different degrees of exhaustion, one being at atmospheric pressure, and the others at the pressures at which the more remarkable effects are observed, and to which we have referred. The effects, which were very interesting, were made apparent to the audience by an image of the instrument under experiment being projected upon a screen by means of a sciopticon lantern.

The influence of the residual gas upon the radiation of heat was beautifully illustrated in these experiments by the greater or less degree of incandescence of the platinum ring with the same battery power at different degrees of exhaustion of the gas surrounding it. The same battery power which would almost fuse the wire in the bulb at higher degrees of exhaustion was capable of only just reddening it at atmospheric pressure, so that, in order to obtain the same temperature of the platinum wire in vacua of different degrees of attenuation, it is necessary to vary the battery power, or what produces the same result, to interpose in the circuit greater resistance as the vacuum becomes more perfect. The comparative temperatures of the rings at different exhaustions were very remarkably illustrated when three instruments, differently exhausted, were joined up in the same circuit, so that the electric current had absolutely the same strength in each case. When this was done the ring in the normal atmospheric pressure was black, that at a higher exhaustion was just red, *while the coil in the still more attenuated medium glowed with incandescence like a small electric lamp.*

Edison appears to have continued his experiments with the lamp described in this patent during the summer and early autumn of 1879. Very little information respecting his work found its way into the newspapers during this period. The unwonted silence seems to indicate that the results of the experiments may not have been altogether satisfactory. A reasonable conjecture might be hazarded that the results of the computations of Mr. Upton had shown that it was impracticable to produce the electric light by the incandescence of metals of the platinum group, at a cost which would permit it to be placed in commercial competition with gas, or at a selling price which the ordinary consumer could afford under any circumstances to pay for it. This was a result, which, as we have seen, had been confidently predicted by Sawyer from his knowledge of the results of Maxim's experiments in 1878,[1] and, indeed, might have been deduced from the

[1] See p. 23, *ante.*

figures given many years earlier by Dr. J. W. Draper.[1] The public mind during the year 1879 gradually became more and more skeptical as to the ultimate success of electric lighting by incandescence. The promises of its promoters had been very profuse, while the performance had been practically nothing.

THE PROBLEM OF THE INCANDESCENT LAMP.

It was clearly understood by the few persons who had an intelligent conception of the true nature of the difficulties yet to be surmounted, that the one thing which now remained to be discovered was the making of an incandescent conductor of material which could withstand without disintegration a much higher temperature than platinum or iridium, and accordingly, the minds of the investigators who were at work in the field were naturally turned in that direction.

As we now know, the final solution of the problem of incandescent electric lighting was found in the substitution for the metallic illuminant in the transparent chamber, of *a conductor formed of carbonized structural or cellular material, in an arch or horse-shoe form.*

LATER WORK OF SAWYER AND MAN.

During the year 1878, Sawyer and Man had pursued with unremitting diligence their experiments and researches in the art of electric lighting. Their first experiment was made about the middle of February, at which time a piece of gas-retort carbon was heated to incandescence by an electric current in a Florence flask, through which a stream of ordinary illuminating gas was kept flowing. This atmosphere was employed because it contained no oxygen to unite with and destroy the carbon while in an incandescent state. An experiment made on the 6th of March, in which paper was accidentally carbonized while being used as a convenient receptacle for a fine line of powdered graphite, first suggested that this material might perhaps be made into a carbon useful in incandescent electric lighting. Following up this experiment, the inventors subsequently succeeded in carbonizing the paper

---

[1] For account of Prof. Draper's experiments and deductions, see *Silliman's Journal* (2d series) iv., 1847. Also *Phil. Mag.* May, 1847.

Professor John Tyndall, in his *Heat as a Mode of Motion*, (4th Ed., pp. 428-9) demonstrated that it was useless to look for an *economical* source of illumination in incandescent platinum, as the proportion of its obscure to its luminous heat rays is as 23 to 1, while from carbon, the proportion is only as 9 to 1, a difference of more than 2 to 1 in favor of carbon.

See also p. 23, *ante*, (Note).

without destroying its structure, by causing a powerful current from a dynamo-electric machine to traverse it while surrounded by an inert atmosphere of hydro-carbon gas.

Although Sawyer and Man were not wholly unsuccessful in carbonizing paper in this way, it was found that the shrinkage of the paper during carbonization caused the strips to break. In endeavoring to overcome this difficulty they formed the strips into the shape of an *arch or horseshoe*, and with this precaution succeeded in making practically useful carbons of paper. But this process of carbonization was found to be not only expensive, but ill-adapted to produce uniformly good results.

During the same month Man enclosed strips of paper with powdered graphite in a closed iron vessel and carbonized them by subjecting them to a high temperature in the fire of his kitchen range in Brooklyn. The carbons were afterwards placed in flasks charged with hydro-carbon gas and brought to incandescence by an electric current. These experiments satisfied the inventors that a good conductor for an incandescent lamp might be made of carbonized fibrous material, as carbons of this description were maintained in a state of high incandescence for two or three hours continuously. From this time forward, Sawyer and Man were continually experimenting in the manufacture of lamp conductors, both of structural and hard carbon, and in the construction of incandescent lamps. Paper and wood were the principal materials carbonized at this time ; many kinds of paper and many varieties of wood appear to have been used. The conductors were of various shapes. Some were straight, some in the form of a flat arch, some in the form of rings with the lower part broken away, and others in the form of an inverted V.[1]

A notice of the exhibition of one of these lamps was published in the New York *Times* of March 19, 1878, as follows :—

A new electric light, adapted to the illumination of dwellings, was exhibited yesterday afternoon at No. 43 Centre street. The lamp is very simply constructed of two metal or carbon rods, between which, near their tops, is held a small bit of compound, pointed at either end and oval in the centre. The components of this compound are the inventor's secret. It is as hard as a diamond, but when subjected to the electric current becomes incan-

[1] See evidence of Albon Man and of W. H. Church in Interference Record. *Sawyer & Man v. Edison.*

descent in a moment. If burned in the open air this compound is rapidly consumed. For this reason the lamp is inclosed in an air-tight glass shade, in which there is a prepared atmosphere. It is said that the bit of compound and the prepared atmosphere will last an indefinite period ; and the cost of the light in dwellings is estimated at one-fifteenth that of ordinary coal gas.

Most of the conductors, before being mounted in the lamps, were subjected to a process which has since come to be variously known as " flashing," "re-carbonizing" and " hydro-carbon treatment." This process was a happy conception growing out of the results of an early experiment above referred to, in which a conductor was maintained in a state of incandescence in an atmosphere of illuminating gas. It was observed by the inventors, that when a carbon was thus treated in a bath of hydro-carbon gas or liquid, its conductivity could be rendered uniform throughout every portion of its length. A deposit of carbon occurs first at the points of highest resistance where the greatest heat is developed, and next upon those of intermediate resistance, after the higher resistances have been reduced by the deposit, and if the operation is continued long enough the deposit may be made to extend over the entire carbon.[1]

Many experiments were made during this period with reference to the most suitable atmosphere for the lamps. It was found that in using ordinary illuminating gas taken directly from the mains, the rapid deposit of carbon soon prevented the incandescence of the conductor, but that when the lamp-chamber was exhausted by an air-pump, the incandescence continued much longer. Hydrogen gas was first tried and found to be in some respects advantageous, but the best results were attained with nitrogen gas. By the first of June, Sawyer and Man had constructed incandescent lamps capable of burning for many days. One lamp was run for many weeks at a luminosity of from 100 to 200 candle-power, and was finally broken by an accident.[2]   On October 15, 1879, the laboratory was removed to a shop at the corner of Walker and Elm street, New York city, at which place the paper and wood carbons were perfected, being mounted in lamps and exhibited to a great number of people. All the lamps made

---

[1] For a fuller description of this invention, see patent No. 211,262, to W. E. Sawyer and Albon Man, dated Jan. 7, 1879 ; also evidence of Albon Man in Interference Record, ans. to Q. 17.

[2] Evidence of W. H. Church in Interference Record *Weston* v. *Sawyer* and *Man*. Nov. 24, 1882. Q. 122.

and used by Sawyer and Man during this time, with the exception of two or three, had enclosing globes wholly of glass. Some of them had globes made of glass in one piece.

Reference has been made to disagreements among the trustees of the Electro-Dynamic Light Company which was formed in 1878 to introduce and place upon the market the inventions of Sawyer and Man. These disagreements were largely due to some of the personal characteristics of Sawyer. He was a man of unquestionable originality and mechanical genius, and possessed a knowledge of electrical science, which although unsystematic and imperfect, was quite extensive, particularly in certain directions. But his habits were in every way bad; he wasted in dissipation almost every dollar which came into his hands, with results which were in every way unfortunate. From the time of the organization of the Electro-Dynamic Light Company, the intemperate habits of Sawyer rapidly gained the mastery over him, so that in March 1879, the officers and trustees of the company were forced to the conclusion that it would be unwise and inexpedient to further continue their efforts to bring out these inventions under the management of Sawyer, who had by this time become wholly unfit to take charge of such important interests. They accordingly deposed him from his position and refused to provide funds for continuing the business. By an arrangement subsequently made, the manufacturing work was transferred to the shops of Wallace & Sons, at Ansonia, Conn., where Sawyer was employed for a time, but his conduct becoming more and more unendurable, he was ultimately dismissed from the works. Angered at this action, he resigned from the directory of the Electro-Dynamic Company in the fall of 1879, and became bitterly hostile to it. He organized a concern under the name of the Eastern Electric Manufacturing Company and thereafter by every effort in his power endeavored to embarrass and break down the Electro-Dynamic Company, and to destroy the value of its property.

The following is a contemporaneous account of the first exhibition of the Sawyer and Man system, from the New York *Evening Post*, of October 30, 1878 :

Some interesting experiments in lighting by electricity were made yesterday afternoon at Elm and Walker streets by the Electro-Dynamic Light Company. The apparatus consists of a small pencil of carbon connected by wires with an electric machine and enclosed in a hermetically-sealed glass globe filled with pure nitrogen gas. The pencil of carbon is heated by the electric current to a temperature of 30,000 to 50,000 degrees Fahrenheit, in an atmosphere with which it cannot chemically combine. The carbon is practically indestructible, and the light is therefore produced without any consumption of material.

Five of these lights were placed yesterday in a darkened room and connected with an electric machine. By the simple turning of a key the light was increased or diminished at will, and when the full current was on the light was equal to about six gas jets, and the volume can, it is said, be made to equal that of thirty gas jets.

W. E. Sawyer of this city and Albon Man of Brooklyn are the inventors of the light. A company has been formed to establish a number of central stations with powerful generators to supply the light to consumers, and it is said that a meter has been invented to show the number of hours each burner is lighted. It is asserted that the cost of the new light is only one-fortieth of the price of gas, and that the necessary connections are easily made.

This was followed by an illustrated description of the system, which appeared in the *Scientific American* of December 7, 1878, from which the following extracts are made :

THE SAWYER-MAN ELECTRIC LAMP.

The practical usefulness of the electric light for illuminating open spaces and wide areas has been amply demonstrated by the various devices for using the electric arc already widely employed. Hitherto, however, it has not been found economical, or even possible, as we understand it, to construct a lamp or candle, based on the electric arc, that would answer the requirements of ordinary domestic and industrial lighting, where a moderate amount of light, well distributed, easily manageable, and of perfect steadiness and softness, is needed. The electric arc seems, from its very nature, to present insuperable obstacles to the economical production of a large number of small lights in a circuit ; in other words, such lights as we require in our dwellings, offices, factories, shops, and the like. And if there were no other means of obtaining light from electricity, the probability of the displacement of gas by it, for the purposes of general illumination, would hardly be worth considering.

The production of light through the incandescence of a pencil of carbon or metal, forming part of an electric circuit and highly heated by its internal resistance to the passage of the electric current, offers an entirely different field for exploration ; and though it has long been apparently closed by the failure of early attempts to obtain an electric light by such means, the achieved success of Messrs. Sawyer and Man, not to speak of the reported success of Mr. Edison, clearly indicates that this is the line along which the practical solution of the problem of household illumination by electricity is to come. The lamp, to be described further on, lacks only the practical demonstration of its economy by protracted use on a large scale, to compel acceptance as a successful solution of the problem.   .   .   .   .   . The adaptability of this form of electric lighting to the needs of household illumination is indicated in Fig. 1. The light produced is pure, strong, and yet soft, like sunlight. It is, moreover, steady and cool. It is not influenced by air currents ; and it does not vitiate the air by poisonous products of combustion, nor by withdrawing the vitalizing oxygen. The lamp takes up less room than the glass shade of a gas jet, and no more than the chimney of an oil lamp. To a limited extent, also, it is portable, and may be used as a drop light. The general appearance of the lamp is shown in Fig. 2. The light is produced by the incandescence of the slender pencil of carbon placed as shown in the engraving. The light-giving apparatus is separated from the lower part of the lamp by three diaphragms, to shut off downward heat radiation. The copper standards lower down are so shaped as to have great radiating surface, so that

FIGURE 1

the conduction of heat downward to the mechanism of the base is wholly prevented. The structure of the base, full-size, is shown in Fig. 3. No detailed description of this portion will be required, further than to say that the electric current enters from below, follows the line of metallic conduction to the "burner," as shown by the arrows, thence downward, on the other side, connecting with the return circuit. The light-producing portion is, of course, completely insulated, and also sealed at the base, gas tight.

The wires supplying the current may be run through existing gas pipes, each lamp being provided with a switch placed conveniently in the wall ; and by simply turning a key the light is turned up or down, off or on. So long as the house is connected with the main it makes no difference to the producer whether all the lights are on or off, since the resistance of the entire (house) circuit must be overcome ; though it will to the consumer, since a meter records the time that each lamp is on, and the charge is rated accordingly. If the Dynamo-Electric Light Company can supply the illuminating force so

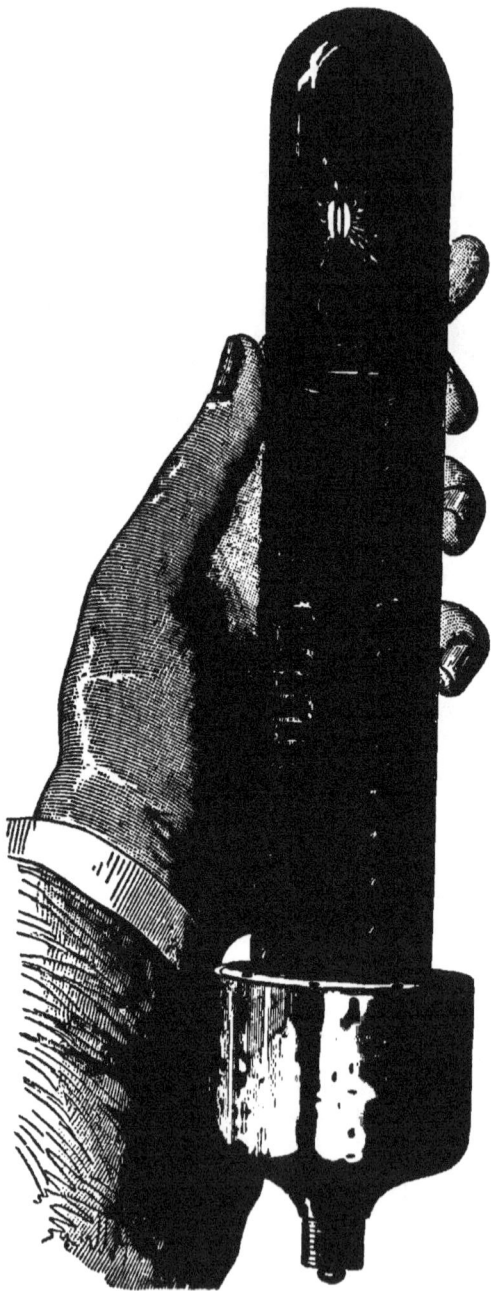

FIGURE 2.

cheaply that the constant and brilliant illumination of all the rooms of a house can be secured at no greater cost than the partial and intermittent

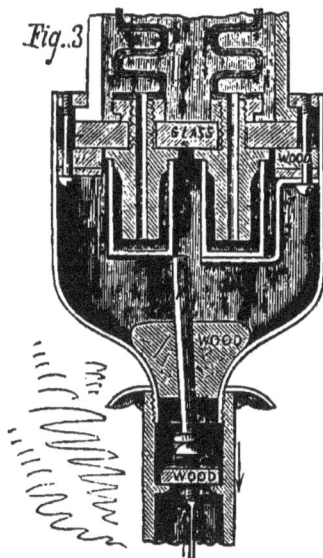

Fig. 3

illumination now had from gas, it is obvious that the electric light will score an important point. The cost of lamps and switches, it is claimed, will not exceed that of gas fixtures.

THE PAPER CARBON CLAIMED FOR EDISON.

While the affairs of the Electro-Dynamic Company were in the unsatisfactory condition referred to ; a condition which necessarily put an end for the time being to all progress in the way of introducing into commercial use the lamps invented by Sawyer and Man, the following announcement appeared in one of the daily journals :—

MENLO PARK, N. J., Dec. 14, 1879.—Extensive preparations have been and are now being made toward the illumination of the streets and private residences at this place on Christmas eve by electricity, under the supervision of Mr. Edison.

The arrangements are rapidly approaching completion, and Mr. Edison is confident that nothing will mar the success of the undertaking. From the number of burners being placed in position, the village will be made as light as day, and the effect can doubtless be seen from a great distance. The event will attract crowds of spectators from among the curious as well as those who are both directly and otherwise interested in the success of Mr. Edison's inventions.

This notice was the forerunner of a somewhat sensational article which appeared in the New York *Herald,* of December

21, 1879, containing an elaborately written history of Edison's
doings in reference to electric-lighting, up to and including the
making of what is now known as the Edison commercial
lamp. As this was the first authoritative announcement of
the abandonment of the original invention, of which so much
had been predicted and expected, and the substitution of an
entirely different material in the incandescent conductor, the
vital feature of the whole system, it attracted no small amount
of attention. The following extracts embody the essential
portions of this publication :

EDISON'S LIGHT.

Tⁿᴱ Gʀᴇᴀᴛ Iɴᴠᴇɴᴛᴏʀ's Tʀɪᴜᴍᴘʜ ɪɴ Eʟᴇᴄᴛʀɪᴏ Iʟʟᴜᴍɪɴᴀᴛɪᴏɴ.

A ꜱᴄʀᴀᴘ ᴏꜰ ᴘᴀᴘᴇʀ.

It Makes a Light, Without Gas or Flame, Cheaper than Oil.

Transformed in the Furnace.

ᴏᴏᴍᴘʟᴇᴛᴇ ᴅᴇᴛᴀɪʟꜱ ᴏꜰ ᴛʜᴇ ᴘᴇʀꜰᴇᴄᴛᴇᴅ ᴄᴀʀʙᴏɴ ʟᴀᴍᴘ.

Fifteen Months of Toil.

Story of His Tireless Experiments with Lamps, Burners and Generators.

ꜱᴜᴏᴏᴇꜱꜱ ɪɴ ᴀ ᴏᴏᴛᴛᴏɴ ᴛʜʀᴇᴀᴅ.

The near approach of the first public exhibition of Edison's long-looked
for electric light, announced to take place on New Year's eve at Menlo Park,
on which occasion that place will be illuminated with the new light, has
revived public interest in the great inventor's work, and throughout the
civilized world, scientists and people generally are anxiously awaiting the
result. From the beginning of his experiments in electric lighting to the
present time Mr. Edison has kept his laboratory guardedly closed and no
authoritative account (except that published in the *Herald* some months ago
relating to his first patent) of any of the important steps of his progress has
been made public—a course of procedure the inventor found absolutely
necessary for his own protection. The *Herald* is now, however, enabled to
present to its readers a full and accurate account of his work from its inception
to its completion.

A LIGHTED PAPER.

Edison's electric light, incredible as it may appear, is produced from a
little piece of paper—a tiny strip of paper that a breath would blow away.
Through this little strip of paper is passed an electric current, and the result
is a bright, beautiful light, like the mellow sunset of an Italian autumn.
    " But paper instantly burns, even under the trifling heat of a tallow
candle !" exclaims the sceptic, "and how, then, can it withstand the fierce
heat of an electric current ?" Very true, but Edison makes the little piece of
paper more infusible than platinum, more durable than granite. And this

involves no complicated process. The paper is merely baked in an oven until all its elements have passed away except its carbon framework. The latter is then placed in a glass globe connected with the wires leading to the electricity producing machine, and the air exhausted from the globe. Then the apparatus is ready to give out a light that produces no deleterious gases, no smoke, no offensive odors—a light without flame, without danger, requiring no matches to ignite, giving out but little heat, vitiating no air, and free from all flickering ; a light that is a little globe of sunshine, a veritable Aladdin's lamp. And this light, the inventor claims, can be produced cheaper than that from the cheapest oil. Were it not for the phonograph, the quadruplex telegraph, the telephone and various other remarkable productions of the great inventor, the world might well hesitate to accept his assurance that such a beneficent result had been obtained, but, as it is, his past achievments in science are sufficient guarantee that his claims are not without foundation, even though for months past the press of Europe and America has teemed with dissertations and expositions from learned scientists ridiculing Edison and showing that it was impossible for him to achieve that which he has undertaken.

HIS FIRST ATTENTION TO ELECTRIC LIGHTING.

When Edison began his experiments in September, 1878, he had just returned from the inspiring scenery of the Rocky Mountains, where he had been enjoying a little recreation after several months of hard labor. He was ripe for fields and enterprises new. A visit to a Connecticut factory where an electric light was used concentrated his thoughts on the subject of lighting by electricity, and he determined to attack the problem. Previous to this time, although he had roamed broadcast over the domain of electricity, wresting from it as is well known, many of its hidden secrets, Edison had scarcely thought of the subtle fluid in connection with practical illumination. Now, however, he bent all his energies on the subject, and was soon deep in the bewildering intracacies of subdivision, magneto currents, resistance laws and the various other branches going to make up a system of lighting by electricity. The task before the young inventor was divisible into two parts.

First—The producing of a pure, steady and reliable light from electricity ; and

Second—Producing it so cheaply that it could compete with gas for general illumination.

HE CHOOSES INCANDESCENCE.

Of the two systems before him, viz.: voltaic arc and the incandescence system, Edison chose the latter as his field of operations. Prominent among the difficulties incident to incandescent lighting, it will be remembered, was the liability of the platinum (when that metal was used) to melt under the intense heat of the electric current, and the liability of the carbon, when that was employed, to gradually become dissipated under the combined action of gases and the electric current.

THE PLATINUM LIGHT.

As between platinum and carbon as the substance to be made incandescent, Edison took up platinum and devoted first his attention to the obtaining of some device to prevent the platinum from melting under the intense heat of the electric current.

The article then proceeds to give a detailed account of the experiments of Edison with platinum lamps and thermal regulators, which, having already been referred to *in extenso* in these pages need not be repeated. Referring to the platinum lamp in its ultimate form the writer of the article says :

The inventor contemplated with much gratification the conclusion of his labors. One by one he had overcome the many difficulties that lay in his path. He had brought up platinum as a substance for illumination from a state of comparative worthlessness to one well nigh perfection. He had succeeded, by a curious combination and improvement in air pumps, in obtaining a vacuum of nearly one millionth of an atmosphere, and he had perfected a generator or electricity producing machine (for all the time he had been working at lamps he was also experimenting in magneto-electric machines) that gave out some ninety per cent in electricity of the energy it received from the driving engine. In a word, all the serious obstacles toward the success of incandescent electric lighting, he believed, had melted away, and there remained but a comparatively few minor details to be arranged before his laboratory was to be thrown open for public inspection and the light given to the world for better or for worse.

A GREAT DISCOVERY.

There occurred, however, at this juncture a discovery that materially changed the system and gave a rapid stride toward the perfect electric lamp. Sitting one night in his laboratory reflecting on some of the unfinished details, Edison began abstractedly rolling between his fingers a piece of compressed lampblack mixed with tar for use in his telephone. For several minutes his thoughts continued far away, his fingers in the meantime mechanically rolling out the little piece of tarred lampblack until it had become a slender filament. Happening to glance at it, the idea occurred to him that it might give good result as a burner if made incandescent. A few minutes later the experiment was tried, and, to the inventor's gratification, satisfactory, although not surprising results were obtained. Further experiments were made with altered forms and composition of the substance, each experiment demonstrating that at last the invention was upon the right track.

A COTTON THREAD.

A spool of cotton thread lay upon the table in the laboratory. The inventor cut off a small piece, put it in a groove between two clamps of iron and placed the latter in the furnace. The satisfactory light obtained from the tarred lampblack had convinced him that filaments of carbon of a texture not previously used in electric lighting were the hidden agents to make a thorough success of incandescent lighting, and it was with this view that he sought to test the carbon remains of a cotton thread. At the expiration of an hour he removed the iron mould containing the thread from the furnace and took out the delicate carbon framework of the thread—all that was left of it after its fiery ordeal. This slender filament he placed in a globe and connected it with the wires leading to the machine generating the electric current. Then he extracted the air from the globe and turned on the electricity.

Presto ! A beautiful light greeted his eyes. He turns on more current expecting the fragile filament instantly to fuse ; but no, the only change is a more brilliant light. He turns on more current, and still more, but the delicate thread remains entire. Then, with characteristic impetuousity and wondering and marvelling at the strength of the little filament, he turns on the

full power of his machine and eagerly watches the consequence. For a minute
or more the tender thread seems to struggle with the intense heat passing
through it—heat that would melt the diamond itself—then at last it succumbs
and all is darkness. The powerful current had broken it in twain, but not
before it had emitted a light of several gas jets. Eagerly the inventor hastened
to examine under a microscope this curious filament, apparently so delicate,
but in reality much more infusible than platinum, so long considered one of
the most infusible of metals. The microscope showed the surface of the
filament to be highly polished and its parts interwoven with each other.

THE PAPER LIGHT.

It was also noticed that the filament had obtained a remarkable degree of
hardness compared with its fragile character before it was subjected to the
action of the current. Night and day, with scarcely rest enough to eat a
hearty meal or catch a brief repose, the inventor kept up his experiments, and
**from carbonizing pieces of thread he went to splinters of wood, straw, paper and
many other substances never before used for that purpose.** The result of his
experiments showed that the substance best adapted for carbonization and the
giving out of incandescent light, was *paper* preferably thick like cardboard,
but giving good results even when very thin. The beautiful character of the
illumination and the steadiness, reliability and non-fusibility of the carbon
filament were not the only elements incident to the new discovery that brought
joy to the heart of Edison. *There was a further element—not the less necessary
because of its being hidden—the element of a proper and uniform resistance to the
passage of the electric current.*

The inventor's efforts to obtain this element had been by far the most
laborious of any in the history of his work from the time he undertook the
task, and without it absolute success to electric incandescent illumination
could not be predicted, even though all the other necessary properties were
present in the fullest degree.

Passing over the scores of experiments made since the discovery that the
carbon framework of a little piece of paper or thread was the best substance
possible for incandescent lighting, we come to consider the way in which the
same is prepared at the present time in the laboratory.

MAKING THE PAPER CARBON.

With a suitable punch there is cut from a piece of " Bristol" cardboard
a strip of the same in the form of a miniature horseshoe, about two inches in
length and one-eighth of an inch in width. A number of these strips are
laid flatwise in a wrought-iron mould about the size of the hand and separated
from each other by tissue paper. The mould is then covered and placed in an
oven, where it is gradually raised to a temperature of about six hundred
degrees Fahrenheit. This allows the volatile portions of the paper to pass
away. The mould is then placed in a furnace and heated almost to white
heat, and then removed and allowed to cool gradually. On opening the mould
the charred remains of the little horseshoe cardboard are found. It must be
taken out with the greatest care, else it will fall to pieces. After being
removed from the mould it is placed in a little globe and attached to the wires
leading to the generating machine. The globe is then connected with an air
pump, and the latter is at once set to work extracting the air. After the air
has been extracted the globe is sealed, and the lamp is ready for use. Figure
7 shows the lamp complete :—

A is a glass globe, from which the air has been abstracted, resting on a stand B. F is the little carbon filament connected by fine platinum wires, G G', to the wires E E', leading to the screw posts, D D', and thence to the generating machine. The current, entering at D, passes up the wire E to the platinum clamp G ; thence through the carbon filament F to G', down the wire E' to the screw post' ; thence to the generating machine. It will be noticed, by reference to the complete lamp in Figure 7, that it has no complex regulating apparatus, such as characterized the inventor's earlier labors. All the work he did in regulators was practically wasted, for he has lately realized that they were not at all necessary—no more than a fifth wheel to a coach.

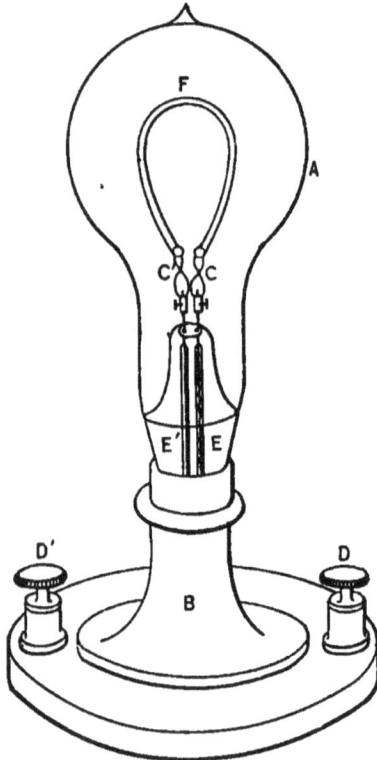

REGULATED AT THE MAIN LIKE GAS—OHEAP.

He finds that the electricity can be regulated with entire reliability at the central station, just as the pressure of gas is now regulated. By his system of connecting the wires the extinguishment of certain of the burners affects the others no more than the extinguishment of the same number of gas burners affects those drawing the supply from the same main. The simplicity of the completed lamp seems certainly to have arrived at the highest point, and Edison asserts that it is scarcely possible to simplify it more. The entire cost of constructing them is not more than twenty-five cents.

EASY METAMORPHOSIS.

The lamp shown in Figure 7 is a table lamp. For chandeliers it would consist of only the vacuum globe and the carbon filament attached to the chandelier and connected to the wires leading to the generating machine in a central station, perhaps a half mile away, the wires being run through the gas pipes, so that in reality the only change necessary to turn a gas jet into an electric lamp is to run the wires through the gas pipe, take off the jet and screw the electric lamp in the latter's place. Although the plans have been fully consummated for general illumination, the outline of the probable system to be adopted is the locating of a central station in large cities in such a manner that each station will supply an area of about one-third of a mile. In each station there will be, it is contemplated, one or two engines of immense power, which will drive several generating machines, each generating machine supplying about fifty lamps.

This article was accompanied by an editorial notice in the

same number of the *Herald*, which, as a matter of contemporaneous comment it may be worth while to place upon record. It is as follows :

EDISON'S EUREKA—THE ELECTRIC LIGHT AT LAST.

In to-day's *Herald* we lay before the public in all its details and completeness what seems to us to be an absolute solution of one of the most important problems that has fixed the attention of scientific men and inventors in our time - the perfection of the electric light. How shall cities, their streets, public places, churches, theatres and public halls be lighted effectively, agreeably, safely and cheaply ? This is an inquiry of dominant importance in municipal administration the world over. And how shall houses be lighted ; how the people in their homes shall weather the difficult point of gas, astral oil, gasoline and similar commodities, is one of the grave issues of domestic economy. For a time—for many months together—the people had their hopes excited on this point, and were led to contemplate the practical application of the electric light to ordinary needs as a boon almost within reach ; as a conquest of science not merely possible but so nearly made that it might almost be counted as in the number of the wonderful additions that the progress of discovery has made to the physical comforts of life. **And then there was a silence about the wonderful invention** and the chronicle bid fair to end like the history of a shooting star.

Everybody was disappointed. Everybody who watches with interest that material progress in the unessential advantages of life upon which depends a great part of what we call civilization, counted the lapse and temporary disappearance of the electric light as an actual loss. It had promised the greatest single advance ever made in the contrivance of artificial methods of illumination, and the failure of the promise left us with a grievous disappointment. In virtue of the glimpses shown in electrical experiments of how near we might be to having in every house these little white stars of the " ethereal essence uncreate," " heaven's first born " of the phenomena of nature, we had come to class all lower artificial lights with the tallow dip and to regard them with the disgust appropriate to unsatisfactory agencies that we had been compelled to cast aside. In fact, all the progress ever made in the improvement of artificial lights before the electric light was extremely little, and to merge the yellow and white gaslight with the white and yellow flame of the tallow dip was not an extravagant hyperbole of disappointment. The dip itself was a noble advancement of human invention ; but the use of gaslight was taught to man by nature in volcanic countries where, through crevices in the rocky ribs of the earth, the escaping vapors caught fire. Perhaps there were exceptions to the number of those disappointed by the apparent failure of efforts to utilize electricity in illumination. All persons deeply interested in the welfare of gas companies stood this failure with great fortitude. Panic had seized upon them at the first reports, and the shares of gas companies had shown the common opinion of what were likely to be the financial consequences of success with electricity. But now they were able to contemplate what seemed the futile labors of E lison with self satisfied complacency. His lamp was even classed already as a " philosophical toy." " Philosophical toy " is the name that has been given in turn to a number of great inventions before the element of cost in them was brought to bear a satisfactory relation to their practical value.

But the story told in our columns to-day will reassure the public whose faith in the " Wizard of Menlo Park " had grown feeble ; will revive the general confidence in the future of electricity as the common agent of domestic and public illumination, and may possibly provoke anew the concern of gas-makers. By this story it will be seen that Mr. Edison has finally elaborated a lamp for the use of electricity that is simpler than any lamp in common use in the houses of the people ; as simple as the gas-burner itself and more manage-able ; a lamp that cannot leak and fill the house with vile odors of combustible vapors, that cannot explode and that does not need to be filed or trimmed. Once more, therefore, the public may reasonably anticipate a time when they will be free from nearly all the annoyances and grievances of ordinary lighting apparatuses and in the full enjoyment, besides, of a light compared to which every other, save daylight itself, is a mere glimmering and gloaming ; compared to which, indeed, even the daylight of about one-third the year is a deterio-rated and adulterated article. People generally knew the soft glory of the light electricity would make, but they never dreamed of the possibility that it could be applied without an apparatus so complicated that it would need a special education to enable them to take care of it.

Perhaps, therefore, this may be regarded as the supreme point of Mr. Edison's achievement in this field – that his last invented lamp is one of such unexpected and remarkable simplicity. His lamp may be screwed on at the end of any ordinary gas burner and the wires from the electric battery may be conveyed to it in the tube that served to convey the gas to the same burner. Whether the wires thus connected come from a battery in another part of the house, or from some larger common reservoir of electricity outside of the house, is a point as to which householders will be able to choose for them-selves, since both sources are contemplated by the inventor. Mr. Edison, it will be seen in the narrative, has contrived a battery for household use which can be adapted to any different number of lamps and to other uses also – can light the house at night and run the sewing machine or rock the cradle all day. For country houses, or for the houses in the city of people who want to keep themselves out of the clutches of corporations that may succeed the gas com-panies, that will be the chosen source of electricity. But in every city this subtle fluid will be made on a large scale and served to houses at a fixed price as gas is now served, and of course the street lamps must be served from such great public laboratories.

Edison's discovery of a substance upon which electricity could produce the light of incandescence without comparative inexpensiveness and perfect effect is one of those little romances of science with which the pathway to every great invention is strewed. Platinum was a great obstacle for a while in this hunt ; and, not altogether satisfactory in operation while of extremely high value, it seemed at the moment as if it might make the search altogether vain. **But the happy discovery of the uses of a bit of cotton-thread has turned in a moment the whole cur-rent of this story into a fortunate channel, and we are rejoiced to congratulate not** merely Edison, but the people of all civilized nations, upon Edison's success.

In *Scribner's Monthly* of February, 1880, there was pub-lished an elaborate illustrated description of Edison's system of electric lighting written by Francis R. Upton, who, as here-tofore stated, had been employed by him as a mathematician.

Referring to Edison's researches which resulted in the discovery of the adaptability of carbonized fibrous material for the incandescent conductor of the lamp, the author says : [1]

He then used a simple thread, which he found to answer the purpose, though it presented the objection of being fragile, uneven in texture, and unmanageable. This difficulty suggested the use of **charred paper**, cut into a thread-like form *The difficulties apparently so insuperable melted away.* The electric lamp was completed. **A piece of charred paper cut into horse-shoe shape, so delicate that it looked like a fine wire, firmly clamped to the two ends of the conducting and discharging wires so as to form part of the electric circuit, proved to be the long sought combination.** From this a light equal in power to twelve gas jets may be obtained. . . . . . . Mr. Edison has thus succeeded in making a lamp of the simplest imaginable construction, and of materials whose expense is extremely small. The paper costs next to nothing, the glass g obes very little, and the platinum tips of the wires are so small that though the metal used is expensive, their cost is trifling. . . **This lamp is Mr. Edison's main discovery.** [2]

The actual history of the invention of Edison's incandescent lamp, substantially in the form in which it is now known to the world, has been made public, both by his own testimony and that of his laboratory assistants, in the interference proceedings which subsequently arose in the Patent Office between Edison and Sawyer and Man, and in the more recent proceedings in the suit of the Consolidated Electric Light Company against the McKeesport Light Company.

It appears that in the summer of 1876, Edison made certain experiments in carbonizing paper, his intention at that time apparently being to manufacture carbon goods for various purposes. He tested the specific electrical resistance of paper carbons in comparison with that of various metals, and found it to be much greater. [3] In August or September, 1877, he mounted strips of this carbonized paper about one-eighth of an inch wide and two inches long, between clamps and passed

---

[1] The article, as published, was prefaced by an autograph letter in fac-simile, as follows :

EDITOR SCRIBNER'S MONTHLY.

Dear Sir :—I have read the paper by Mr. Francis Upton, and it is the first correct and authoritative account of *my* invention of *the* Electric Light.

Yours truly,

Menlo Park, N. J. . THOMAS A. EDISON.

Although, by what must be charitably viewed as an unfortunate slip of the pen, the personal pronoun and the definite article in the last line became transposed, a circumstance which has given rise to a perhaps unnecessary amount of critical comment, this letter may nevertheless be construed as an official endorsement of the statements contained in Mr. Upton's article

[2] *Scribner's Magazine*, February, 1880. pp. 536–38.

[3] Interference proceedings, *Sawyer & Man* v. *Edison*, 1881. Evidence of T. A. Edison, in answer to Q. 2, 6.

an electric current through them. The carbon was made incandescent, but was quickly oxidized and destroyed. The experiment was then repeated in a partial vacuum produced by a common air-pump, but with little if any improvement in the result.[1] A similar experiment with incandescent platinum has been described. In October, 1879, apparently becoming dissatisfied with the prospects of reaching any results of commercial value in electric lighting by the use of incandescent platinum, he commenced experimenting with carbon. One of his assistants had about a year before this made a great number of carbons of tissue paper, coated with a mixture of lampblack and tar, rolled into the form of a knitting needle and carbonized.[2] It appears that Edison afterwards had these threads of carbon, as he called them, measured electrically, and the resistance was found to be so great as to suggest the possibility of using them in lamps. Edison subsequently testified that it was in September or October, 1879, that he for the first time reached the conviction that electric lighting could be accomplished by the use of carbonized paper in a high and stable vacuum.[3]

The evidence of Charles Batchelor, Édison's principal assistant, as to the history of the incandescent lamp invention is of much interest. Mr. Batchelor says :

" During the last part of the year '78, and up to October, 1879, I made, at Mr. Edison's request, a very large number of lamps having platinum and platinum-iridium composing the incandescent conductor. A great many of these lamps had their conductors coated with insulating material in order to be able to wind them up close and get them into as small a space as possible, in order to offer the least radiating surface. Mr. Edison very frequently sat down at my table and worked for hours helping me on these experiments. Our conversation frequently was directed to getting the highest resistance in the least possible space. I remember once or twice during these conversations, early in 1879, he remarked how easy it would be to get this resistance if carbon was only stable. During the time that I was experimenting on these lamps he had been busy experimenting to perfect the different apparatus composing his electric lighting system as a whole. I had also worked on these matters, but as our lamp was an exceedingly difficult job, the majority of my time, both day and night, with the exception of a week or two in which I devoted some time to telephones, was spent on the lamp. He had succeeded in making a more perfect dynamo-machine. In testing the lamps with platinum conductors he had been continually improving the apparatus for exhausting the globes. In

---

[1] Ibid. Answers to Q. 7, 8.

[2] Ibid. Answers to Cross-Q. 169-187.

[3] Ibid. Answer to Cross-Q. 259. Batchelor says "about October 22, 1879." See p. 47 *post.*

October, 1879, when he had got a very perfect vacuum for his lamps, he suggested the use again of carbonized paper as a conductor, and accordingly he had me cut a fine filament of paper which we carbonized and put in a globe. This filament, I believe, was cut straight from paper and bent round previous to putting in the carbonizing chamber. I do not remember what we did with this lamp afterward; but within a day or so of that I cut a loop from paper similar in shape to the one now in Edison's Commercial Incandescent Electric Lamp. At the same time that these were being tried I also made lamps of loops of carbonized thread, carbonized flax, fine filaments of lampblack and tar rolled up and baked, and also threads which had been treated with lampblack and tar previous to carbonization. All these things were used about the same time as incandescent conductors in electric lamps, the most satisfactory at the time being the carbonized paper loop which I had cut by hand. We immediately after this made a steel mold in which these loops could be cut quickly, and after a few experiments in the carbonization of them, in order to get their resistance as near as possible alike after carbonization, we made a number of these filaments and used them at an exhibition in Mr. Edison's house, about the second or third of December, 1879. When the first lamp was made, which had a fine filamentary carbonized paper conductor from which the light was given, which was about October 22, 1879, then I believe we had a system of electric lighting that was complete and could compete with gas, and we proceeded as expedit ously as possible to exhibit it as such." [1]

It has been claimed that Edison fully understood the advantages of structural carbon for an incandescent conductor long before he succeeded in making such a conductor practically operative, and by way of explanation, it has been said that while he could not at that time avail himself of this superiority, for lack of a vacuum pump to exhaust the chamber of his lamp, he did in fact adopt the structural carbon as soon as he secured a suitable pump. [2] This assumption seems to be negatived by the fact that as early as April, 1879, Edison certainly had in use the combination of the hermetically sealed lamp entirely of glass, the high resistance illuminating conductor, and the high and stable vacuum. [3] Nevertheless, he did not then put the structural conductor into that lamp, but on the contrary, continued to experiment with and use platinum conductors until at least as late as the

---

[1] *Consolidated Electric Light Co.* vs. *McKeesport Light Co.*, 1889. Def't's Record, pp. 314–15.

In his argument at the hearing of *Consolidated Co.* v. *McKeesport Co.* in the U. S. Court in Pittsburgh, May, 1889, Counsellor Grosvenor P. Lowrey said :

It was not until the experiments of physicists, who had obtained a more perfect vacuum, came to the knowledge of Mr. Edison, that he said "here it comes, here is the condition ; now we will dismiss platinum and go back to carbon which we always knew was the thing, but which we could not burn in the open air." . . . . A fine fibre of carbon enclosed in a lamp chamber, from which the air has been exhausted to the millionth of an atmosphere, a thing not known until 1878 or 1879, when it became possible by the adoption of the Sprengel pump.—[Vide report in *Electrical Engineer*, viii, 287, June, 1889.

[3] This combination is described in the specification of patent No. 227,229, see p. 26, *ante*.

following October. It is quite apparent that he did not prior to that time know that any organic carbon conductor could be made sufficiently homogeneous to fulfil the conditions sought for. In fact, no evidence has been brought forward, tending to show that he used carbon of any kind in an incandescent lamp before the fall of 1879.

Edison filed his application for a patent on the paper carbon on December 11, 1879. The specification and claims are as follows:

To ALL WHOM IT MAY CONCERN :

Be it known that I, Thomas Alva Edison, of Menlo Park, in the State of New Jersey, United States of America, electrician, have invented an improvement in electric lamps and in the method of manufacturing the same, of which the following is a specification :

In a former application made by me for letters patent in the United States an improvement in electric lamps is set forth, wherein a filament of carbon is enclosed in a glass bulb, and the atmosphere removed as nearly as possible, and the carbon is brought to incandescence by an electric current to form the lamp.

My present invention relates to an improvement in the process of manufacturing the carbon filament and in the means for securing the same to the conductors.

I make use of paper of the desired thickness as free as possible from fore gn substances or adulterations, and for this purpose I prefer and use " bristol board." With suitable instrument, such as a puuch and die, I cut out a narrow strip of this paper, preferably in the form of an elliptical bow or an arc of a circle, the ends of the strip being by preference wider than the other portions.

A number of these pieces of paper are laid flatwise in the bottom of a mold preferably of wrought iron, and there is laid on them a light weight in the form of a flat piece of gas retort carbon or otner device that will not be distorted by the heat. If several of these are laid one on the other in the mold, a piece of tissue paper is interposed between each one and the next.

A cover is used to close the mold, and the mold is raised very gradually to a temperature of about six hundred degrees Fahr. This allows the volatile portion of the paper to pass away, and at the same time the mold retains the paper in its proper shape, and the paper is prevented from curling up or becoming distorted, as it would be likely to do if the heat were applied suddenly, or the light weight dispensed with.

The mold is now placed in a furnace and heated almost to a white heat, and then removed and allowed to cool gradually.

The carbon filaments will be found to be smaller than the cardboard blanks, and to be sufficiently strong and flexible for handling. The ends of the carbon are to be secured to the metallic conductors in any convenient manner.

The carbon filaments prepared as aforesaid are very uniform in their resistance to the electric current, and I make them thin and of a sufficient length to offer a great resistance to the passage of the current.

The clamps that connect the conductors to the ends do not require to be pressed with much force on the carbon, because the resistance to the passage of the current between the clamps and the carbon will be less than the resistance of the carbon filament, hence but little heat will be developed at the clamps.

In ordinary electric lamps the large carbons do not offer much resistance to the electric current, and unless the clamps are very firmly pressed upon the carbon. the current meets with considerable resistance at the clamps and hence heat is developed at such clamps.

The clamps that I prefer are made of a steel spring tipped at the ends with platina or similar metal; the spring is bent into a circle or bow, and the ends crossed and turned back toward each other similar in shape to the Fig. 8 with the opening to the carbon between the spring ends of the upper part; the object of this shape is that the pressure of the clamps on the carbon may be increased by the expansion of the spring by the heat of the lamp instead of being lessened, as it would be if the wire was only bent into a single bow.

The spring clamps are connected to the platina or similar conducted wires by clips, and the platina wires pass through the glass of the globe or bulb that contains the lamp. The air is to be exhausted from the bulb by any suitable means, and it is preferable to exhaust said air as perfectly as possible, say to the one eight-hundred thousandth of an atmosphere.

The lamps are suspended or supported in any convenient manner, and the electric current from a magneto electric machine or other source of electricity is passed through the lamp, and brings the carbon filament to a high incandescence and the lamp is very durable, and a large number of such lamps can be placed in the electric circuit in multiple arc, or otherwise, as desired.

In the drawing:—

Fig. 1 is a vertical section of the lamp complete.

Fig. 2 is a side view on large size of the clamping device.

Fig. 3 is a section at the line xx in still large size.

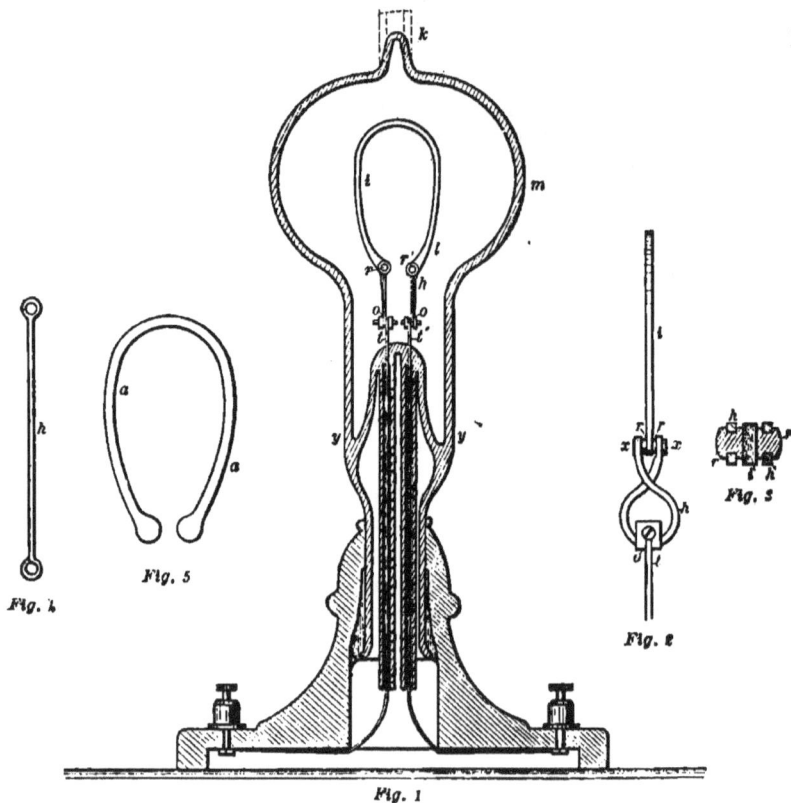

Fig. 4

Fig. 5

Fig. 1

Fig. 2

Fig. 3

Fig. 6

Fig. 4 is the wire forming one of the clamps before it is bent up to shape.

Fig. 5 is the paper blank before it is carbonized, and

Fig. 6 is a section of the box.

The blank a is cut out of paper material such as "bristol board," in the proper shape, the form shown Fig. 5 is preferred, the same is laid in the meta mold *b*, and when several are laid one on the other, pieces of thin paper are introduced between. The weight *d* is laid on these; it is to be heavy enough to prevent the paper curling up under the action of the heat, but it allows the paper to contract as the volatile matters are expelled by the heat. This weight *d* is of gas retort carbon. The cover *e* is placed on the mold and secured, and the mold is heated as before described.

The carbon filament *i*, forms the lamp when rendered incandescent by the electric current passed through it.

The clamp is made of the wire *h*, at the ends of which are tips or small rivets *r*, of platina or similar material.

The wire is bent up and crossed as shown, so as to act as a spring in clamping the end of the carbon filament that is placed within such clamp. The wire is attached to a small stock *o*, into which the conducting wire *t*, passes and is clamped.

The conductors for the two ends of the carbon are inserted into the glass and the latter intimately melted around them; the carbon and clamps are connected to the wires and the parts introduced within the neck of the bulb *m*, and the glass melted at *v*, the air is exhausted from the globe by the tube *k*, that passes away as shown by dotted lines and the tube melted together while the vacuum is maintained.

The lamp is ready for the conductors to be attached to it, and the carbon is rendered incandescent by the current that passes through the same. It is durable, as there is nothing to combine with the carbon and it is substantially indestructible.

I claim as my invention:

First. The manufacture of carbons for electric lights from paper.

Second. The method herein specified of manufacturing carbons for electric light, consisting in exposing the filament of paper to the action of heat in a mold to drive off the volatile portions and carbonize the paper, substantially as set forth.

Third. A carbon for electric light made as a filament with the ends broader for the clamping devices that connect the conductors.

Fourth. The clamp for the carbon of an electric lamp composed of a bow or elliptical spring with the ends crossing each other, and receiving between them the carbon substantially as set forth.

Signed by me, this 8th day of December, A. D. 1879.

THOMAS A. EDISON.

COMMENTS OF THE PRESS.

The effect produced by the announcement of Edison's discovery of the carbonized paper lamp was various. Public opinion, however, seems to have largely coincided with that of Fontaine, the well-known French electrician, who, in explanation of the fact that the announcement in the *Herald* was received with so much incredulity by scientific men, said, " I am not surprised that people received with mistrust the announcement often repeated of Edison's discoveries. It is

after the fashion of the shepherd who was always crying
'wolf.'" A very general impression appears to have existed,
which was reflected in many of the leading journals of the
day, that the announcement was designed merely as a basis
for a stock speculation. The following extracts, among many
others which might be cited, are examples of the prevalent
feeling :

> The wonderful electric lamp which Mr. Edison was said so long ago to
> have invented, and for a sight of which the world has been impatiently wait-
> ing, seems to be a failure. At least, according to the *Herald*, Mr. Edison has
> himself abandoned the spiral of platinum and iridium which were its main
> feature, and the ingenious mechanism by which they were kept in a state of
> incandescence without fusion, and has turned for a solution of the problem to
> incandescence carbon. The result—the perfected lamp—appears to be a
> modification of the Sawyer-Man light, which, as everybody knows, consists of
> a fine incandescent pencil of carbon in a globe of nitrogen gas. Mr. Edison
> bends his carbon pencil, or filament, into the form of a horse-shoe and exhausts
> the globe surrounding it. That is substantially the only difference between
> the two systems. Both depend upon a fine thread or cylinder of incandescent
> carbon. Both protect the carbon from the action of oxygen, the one by filling
> up the globe with nitrogen, the other by drawing out the oxygen with an air-
> pump. There is no new principle involved, therefore, in Mr. Edison's light,
> and the modifications he has made in the Sawyer-Man light are but a poor
> return for fifteen month's labor. The new Sawyer-Man-Edison lamp, however,
> will probably prove a useful addition to other forms of electric lighting,
> although it is not likely to be regarded as a complete and satisfactory solution
> of the great problem which has so long been absorbing the attention of elec-
> tricians.—[*N. Y. Tribune*, December 22, 1879.

> A plausible story of success justly alarms the holders of shares in corpo-
> rations which supply gas. On the other hand, the invention of Mr. Edison,
> which is *nil* as yet, is the property of a corporation whose shares can be and
> have been run up to fabulous prices, or rather quotations, by the same story.
> The whole outcome of the Edison light is a double stock-jobbing enterprise.
> Bears in gas stocks and bulls in Edison light stock have had an opportunity to
> reap rich harvests. Nobody else has been benefited, or is likely to be for any-
> thing that appears as yet. We should be slow to accuse Mr. Edison himself
> of using his reputation as an inventor in such business, but if he has any
> sagacious friends they should let him see how his influence is employed in a
> base and unscrupulous manipulation of stocks. The great public has probably
> learned by this time that as yet no great reliance is to be placed on Mr.
> Edison's own opinion of his inventions.—[*Boston Advertiser*, December 23,
> 1879.

> Mr. Edison promises to exhibit his carbonized paper device on New Year's
> eve. There is nothing in the elaborately described process which has been
> published with such eclat, to convince any competent electrician that he has
> any certain prospect of success. Meanwhile agents are publishing the *Herald's*
> description in Western cities, and seeking to organize stock companies to

supersede gas, oil, and all other means of illumination, with this new light, which no one but its inventor has seen.—[*Cleveland Herald,* December 25, 1879.

Mr. Edison ought to know that the public will not be disposed to put its faith in him much longer if he does not take prompt measures for narrowing the gap between his fullness of promises and his lack of performance concerning the electric light. He ought also to be informed that the delay in putting his invention to a test which can be accepted as a searching and impartial test has already affected not only the reputation which he has acquired as a mechanical genius, if not as an instructed man of science and a competent electrician, but also his reputation as an inventor devoted singly to his own pursuit. It is extensively believed that (in the language of lamps) he he has turned the light alternately on and off the progress of his invention in a manner unpleasantly favorable to the manipulation of gas stocks. The holders of gas stocks have not forfeited their rights as citizens by buying gas stocks, and although they would have no right to complain if their investments were suddenly destroyed by the invention of a light better and cheaper than gas, they certainly have a clear right to complain if illusive proclamations of the invention of such a light are so put about that stock-jobbers may make use of them effectively to depress the price of gas stocks.—N. Y. *World,* Jan 5, 1880.

Not only is Mr. Edison to be congratulated on the happy past, but his friends may look forward to a long and equally happy future, crowned at periodical intervals by similar dazzling and final triumphs ; for, if he continues to observe the same strict economy of practical results which has hitherto characterised his efforts in electric lighting, there is no reason why he should not for the next twenty years completely solve the problem of the electric light twice a year, without in any way interfering with its interest or novelty. —[*Saturday Review,* January 10, 1880.

On the other side of the Atlantic, at about the same time, a series of bulletins was put forth, which not unnaturaly gave rise to a similar suspicion. For example, on the 3d of January the *Figaro* published as from "Jersey City, one of the faubourgs of New York," a cable message said to have been sent by Edison's secretary, S. L. Griffin, to this effect :

During the past week five houses have been lighted every night by the new electric lamp of Edison. The success was perfect and was witnessed by an immense crowd. **There is a great fall in gas shares.** The shares of the Edison Electric Company, issued at five hundred francs, are quoted at nearly twenty thousand francs.

The next day, January 4, the following despatch appeared in the London *Times :*

PHILADELPHIA, January 4.—Mr. Edison's carbon horse-shoe electric lamps continued successfully burning on Saturday night at Menlo Park. There were about 100 in position, which were supplied by two generators. Menlo Park was so overrun with visitors and investigators last week that the manufacture of the lamps was impeded. *Scientific criticism of the success of Mr. Edison's in-*

vention has almost ceased in the United States, the public generally regarding the invention as succcesful.

A gentleman in Paris connected with the publication department of *Figaro*, in order to assure himself of the truth of this statement, cabled to Edison and received the following reply, which appeared in that journal on January 6:

> MENLO PARK, January 4, 1880—*Times* telegram correct. Practicability and economy greater than claimed. All houses here lighted ten days.
>   EDISON.

On the same day on which this cablegram was received from Edison, the Paris correspondent of the Edinburgh *Scotsman*, who appears to have seen the above announcement as well as another from Edison's secretary, sent the following despatch to his newspaper in Scotland, in which it duly appeared on January 6:

> PARIS, January 5—Evening.—There seems little doubt that the latest form of electric light devised by Mr. Edison is by far the simplest, and consequently the cheapest yet invented, and it is apparently the only kind that can be utilized for household purposes. Two telegrams, which will be published to-morrow, have been received here—the one from Mr. Edison himself and the other from his secretary—and both confirm the estimate of the *Times'* correspondent. Indeed they both declare that the correspondent rather undervalued than overstated the advantages of the light. Mr. Edison says, 'Practicability and economy greater than claimed. All houses at Menlo Park lighted for the past ten days.' Mr. Griffin, his secretary, adds, 'Fifty-three lights burning publicly over two weeks. Competent scientists pronounce tests perfect.'

Mr. Griffin's bulletin, published by *Figaro* next day in full, is as follows :

> My despatch rather under than over estimated. Fifty-three lights burning publicly over two weeks. **Competent scientists pronounce tests perfect.** Four thousand dollars bid for stock, $5,000 asked.   GRIFFIN, Secretary.

It is but fair to say, however, that the announcements quoted above, in spite of the suspicions which they engendered, did not exaggerate the actual success which had been attained in the experiments at Menlo Park. Thousands of persons visited the exhibition, and the interest and curiosity of the public were excited to a high pitch.

INTERFERENCE PROCEEDINGS INSTITUTED.

It was at once perceived by the Electro-Dynamic Company that the real invention, the fundamental thing which gave value to the new and apparently successful lamp, was *the incandescent conductor of structural carbon, to which had been*

given the form of an arch or horseshoe, the identical combination which had long before been invented and reduced to
practice by Sawyer and Man. Moreover, a full description of
their invention had, early in the autumn of 1878, and long
before the announcement of Edison's new lamp, been placed
in the hands of an attorney for the purposes of securing a
patent, but owing to the disagreements which have been
referred to, as having arisen between Sawyer and his associates in the Electro-Dynamic Light Company, he had
refused to affix his signature to the application. It was not
until after the publication of the account of Edison's invention had been made, that he could be prevailed upon to sign
the application, and conjointly with Man, to assert their prior
right as joint inventors to the invention.

The filing of the application in the Patent Office led to
what is known as an Interference, which is in the nature of a
judicial proceeding, the evidence of witnesses being taken
under oath for the purpose of determining the question of
priority of invention, as between two applicants for a patent,
both of whom claim to have originated the same, or substantially the same invention.

Interference proceedings in the case of *Sawyer & Man v.
Edison* were instituted by the Commissioner of Patents on
September 23, 1880, and were protracted for nearly five years.
Much testimony was taken on behalf of each of the parties,
and the history of the invention in dispute was thoroughly investigated. The case was argued first before the Examiner of
Interferences, who on January 20, 1882, rendered a decision
awarding priority of invention to Sawyer and Man. A motion
was immediately made by Edison to reopen the case to permit the taking of newly-discovered testimony, it being alleged
by him that the admission of such testimony would show that
Sawyer and Man were not legally joint inventors ; that whatever either or both of them had done in respect to the invention in controversy, prior to the filing of their application,
amounted to nothing more than unsuccessful and abandoned
experiments, and that Sawyer himself had publicly given
Edison credit for the invention in issue, claiming for himself
and Man only carbon manufactured from hydro-carbon liquids,
and carbonized willow twigs. Numerous affidavits were put
in in support of these allegations. The charge of abandon-

ment was supported by a number of letters written and published by Sawyer in the newspapers, immediately after the
appearance of the accounts of Edison's lamp in the *New York
Herald*, in December, 1879.[1]

On July 24, 1882, Commissioner Marble rendered a decision setting aside the decision of the Examiner of Interferences,
and reopening the case. An extract from this decision will
sufficiently show the grounds upon which it was predicated.
After quoting at length from various statements published by
Sawyer, he says :

> The quotations above given have a tendency, at least, to show that whatever
> had been done by Sawyer and Man prior to the time said letters were written in
> relation to the invention herein claimed, was simply experimental, and that
> their experiments had fallen short of success Sawyer not only says that their
> efforts had proven ineffective, and were considered by both himself and Man as
> worthless, but that the abandoned institution had been placed in the hands of
> parties where it could be found. In his second letter he very forcibly denies
> the truthfulness of statements made as to the operativeness of Mr. Edison's
> lamps. Notwithstanding these letters were written and published as above
> stated, on the 9th of January, 1880 an application, executed by Sawyer and
> Man, describing and claiming the invention in question, was filed in this office.
>
> In a work published by Mr. Sawyer, and received at this office on March
> 12th, 1881, Mr. Sawyer says on page 72: "The carbonizing live willow twigs,
> with a view to obtaining a suitable bent carbon, by Sawyer and Man, and the
> carbonizing of paper and bamboo by Edison, substantially close the account of
> incandescent carbon." On page 78 he says: "The filament originally used by
> Mr. Edison was prepared by cutting cardboard into the desired shape, and car-
> bonizing the same by placing the loops thus formed in layers within an iron
> box with intervening layers of tissue paper, closing the box to exclude oxygen,
> and raising the whole to red heat in a furnace."
>
> Taking all of the statements of Mr. Sawyer together without further ex-
> planation, I am compelled to reach the conclusion that the experiments made by
> himself and Man were not successful, and that they did not have the invention
> completed on December 24, 1879.   .   .   .   .   .   .
>
> In view of all these facts disclosed by the affidavits filed I think that
> Edison should be permitted to prove, if he can, that Sawyer and Man were not
> joint inventors, and that what experiments were made by them were mere ex-
> periments, and did not show a complete invention.

New testimony having been introduced at great length
under the above order of the Commissioner, the case was re-
decided on June 2, 1883, by the Examiner of Interferences, who
again adjudged Sawyer and Man to be the prior inventors.
This decision, although reversed by the Board of Examin-
ers, was affirmed by the Commissioner. Edison, as the unsuc-

---

[1] Vide p. 38 *ante.*

cessful applicant, then appealed to the Secretary of the Interior, who, on November 19, 1884, dismissed the appeal for the want of jurisdiction. In the meantime, Commissioner of Patents Marble having resigned, and Mr. Butterworth having been appointed in his place, Edison applied for a rehearing. The question was elaborately argued, but on February 14, 1885, Commissioner Butterworth denied the motion, thus leaving the matter to stand as decided by Commissioner Marble.

The following extracts from the decision of Commissioner Marble will serve to show the general considerations upon which the judgment was founded. It should be premised by way of explanation, that it had been argued in behalf of Edison, not only that the respective inventions in the controversy were not the same, but also that Sawyer and Man were not in a legal sense joint inventors of the subject-matter claimed by them. In his decision, the Commissioner said:

Edison uses his carbon in a glass bulb from which the air has been exhausted by a pump to the millionth of an atmosphere. Sawyer and Man use their carbon in a similar bulb from which the air has been expelled by the introduction of nitrogen or hydrogen gas. **Both carbons are used in substantially the same way and for the same purpose.**

It is urged in behalf of Edison that because Sawyer and Man treat the paper before it is carbonized, and subject the carbon after it has been formed to the treatment of hydro-carbon gas, as stated by Mr. Man, it is no longer an incandescent conductor for an electric lamp formed of carbonized paper, but a compound conductor, and that the paper carbon is simply a frame work upon which a coating of carbon is deposited, and that the original character of the paper carbon is entirely changed.

On behalf of Sawyer and Man, it is urged that notwithstanding the treatment to which the paper and carbon are subjected, the carbon is still one formed of carbonized paper. The only witness who testified understandingly as to the effect which the deposit of carbon had upon the paper carbon, was Mr. Batchelor, a witness called in behalf of Edison.   .   .   .   .   .

.   .   From Mr. Batchelor's statement, it will be seen that the paper carbon is not affected by such treatment except by the coating of carbon which it receives. In all other respects it remains the same.

The invention, as claimed, is broadly to the incandescent conductor for an electric lamp, formed of carbonized paper. The claim is without limitation, and was intended to cover, I think, any kind of a conductor for an electric lamp formed of carbonized paper.

Can it be truthfully said that the conductor made and claimed by Sawyer and Man is not a paper carbon? Before its treatment in hydro-carbon vapor it certainly is. Does that treatment change its entire character? Is the desk at which I sit not a wooden desk, because covered with a coating of varnish? Is the paper on which I write not paper, because sizing was used in the manufacture to give it body and finish? Are the doors which lead into this room not wooden doors, because there is a coating of paint on them ? The desk and the doors are undoubtedly improved by the coatings of varnish and paint, because made more durable in any atmosphere, and under any conditions, but they are used in the same manner as if the coatings were not on them, and are known by the same names.

So with Sawyer and Man's carbon ; it is undoubtedly changed somewhat by the coating of carbon deposited thereon ; but it is nevertheless, I think, a paper carbon. The carbons, when prepared by either of the parties in the manner described by them, become incandescent conductors when placed in a lamp from which the oxygen has been removed, either by exhaustion by a pump, or by expulsion by hydrogen or nitrogen gas. Neither will operate unless the oxygen is removed, and both will when it is. In other words, each of the carbons become an incandescent conductor under substantially the same conditions. They are, therefore, substantially the same, and must be so held.   .   .   .   .

In their efforts to carbonize paper, the testimony of Man and Sawyer, as well as other witnesses, shows that both the parties were engaged in devising ways and means to secure a perfect carbon from paper, very likely each of them made suggestions which were adopted; they were working together for a common end and object, viz. : The carbonization of paper, which was finally accomplished in the manner described by Mr. Man.

Where two or more parties thus unite their efforts to accomplish a particular object, and by such efforts that object is secured, I think it must be held that the product obtained, if any is obtained, is the joint product of both. There is no doubt in my mind that Sawyer and Man should be considered joint inventors of the incandescent conductor made by them.

The case, therefore, recurs on the question of priority of invention.

After quoting at length from the testimony in the case, the Commissioner sums up as follows :

I have thus given sufficient of the testimony of the respective parties to this contest to show what they did respectively, and when they did it.

Edison completed the invention and reduced it to practice on or about October 22, 1879. Sawyer and Man had the invention in a rude form as early as March, 1878, and completed as early as September or October of the same year. *The question here is, not when was the invention so completed as to compete commercially with gas, but when was it completed as an operative invention?* It may be that it will never prove to be equal to gas commercially, but that will not prove it is not a complete and perfected invention. *An invention is complete when the thought conceived is embodied in some practical and operative form.* It may never prove to be profitable because of other ways devised for accomplishing the same end, *This invention was complete in a legal sense when either of the parties to this controversy had devised means by which to demonstrate its practicability.* This was done, not when they with a conductor formed of carbonized paper had produced a light which would continue for a few minutes, but when they with such conductor had produced one which would last for hours, days and even weeks. Such a light with such conductor was first produced by Sawyer and Man.   .   .   .   .

Upon the question of abandonment by reason of the publications and statements made by Sawyer between the date of publication of the first account of Edison's lamp, and the filing in the Patent Office of the application of Sawyer and Man, the Commissioner says : .

Before the writing of these letters and publications the invention of Sawyer and Man had passed beyond the tentative domain of experiment, and had reached the solid ground of completion by practical demonstration. If it be conceded, therefore, that Mr. Sawyer wrote the letters attributed to him, and published the book put forth in his name, they cannot be allowed, either singly or together, to work an abandonment of the invention of Sawyer and Man. *This invention was the product of their joint efforts, and could not be abandoned by either, separately and alone.* Mr. Sawyer could not by any act or by any word prejudice the rights of Mr. Man without the consent of the latter. I do not think, however, that it was the intention of Mr. Sawyer to abandon the invention, but, feeling piqued and annoyed by the success of Mr. Edison, he published his letters for the purpose of calling attention to his own and the joint inventions of Sawyer and Man. At the time said letters were written and the book published, he was inimical to the owners of this invention, and his habits for sobriety were not good, to say the least. Under such conditions it is not strange he should have not written and published what, under other circumstances, he would not have uttered.

In view of the facts disclosed in the testimony regularly taken in this case, I do not consider any of the statements made by him in his letter or book of any importance, and hence dismissed them from consideration. Neither do I consider, for the foregoing reasons, the question of where the title of the invention was, at the time said letters were published, of any moment to the decision of this case. The question here is not as to who, at that time or now, owns the invention of either of the parties, but which of the parties to this record first made the invention. I think it is fully and clearly shown that Sawyer and Man were the first inventors of the "incandescent conductor for an electric lamp formed of carbonized paper."

The able counsel of Mr. Edison were by no means willing to rest satisfied with the final result of the interference proceedings, and they at once petitioned for a renewal of the motion for a rehearing. The parties were heard before Acting Commissioner of Patents Dyrenforth on March 12th, 1885, and on April 2nd following a decision was rendered by him denying the motion. As this decision affords a summary review of the proceedings, it is here given in full :

This motion for a rehearing on behalf of Edison, after reciting the leading facts in the proceedings heretofore had, including the fact that Commissioner Marble duly decided against the petitioner on the issue defined in the interference, and the fact that Mr. Commissioner Butterworth duly heard and denied a motion for a rehearing, proceeds to set forth the grounds why a rehearing should now be granted. These grounds are substantially as follows:

That the interference was based on the assumption that the two applications described the same invention, whereas the original application of Sawyer and Man did not disclose the invention in interference, and the testimony taken shows that the said invention was never made by Sawyer and Man, as has always been insisted upon by and on behalf of Edison; that is to say, rehearing is asked on the general allegations that there never has been an interference between the inventions of the parties, and that Edison has always said so, but has never been listened to.

I need scarcely remind counsel that since the entire proceedings, from the declaration of interference to the present time, have been based on the judgment of the office that there was an interference, a general denial of the correctness of that judgment is not ground of reviewing one's own decision, much less the decision of a predecessor. I may also call attention to the remark of Mr. Justice Field, in *Giant Powder Co.* vs. *Vigorite Powder Co.* (5 Fed. Rep., 197), as follows:

" A new hearing should not be had simply to allow a rehash of old arguments," which seems to be pertinent to this case. And I may conclude by quoting the closing paragraph of the decision, dismissing the former petition for rehearing, in which I concur.

The courts are open, and have a broader and fuller jurisdiction than I possess. If there is error in the decision of my predecessor, which works a hardship upon either of the parties to this controversy, there is another forum where his action can be reviewed.

There must be an end of litigation over a case in this office, and as I think that this case reached that end at date of the decision last mentioned, I must decline to re-open it. The motion for a rehearing is therefore denied.

R. G. DYRENFORTH, Acting Commissioner.

This decision was followed in due course by the issue of a patent on the application of Sawyer & Man, which is here given as a matter of record.

UNITED STATES PATENT OFFICE.

WILLIAM EDWARD SAWYER, OF NEW YORK, AND ALBON MAN, OF BROOKLYN, N. Y., ASSIGNORS TO ELECTRO-DYNAMIC LIGHT COMPANY, OF NEW YORK.

ELECTRIC LIGHT.

Specification forming part of Letters Patent No. 817,676, *dated May* 12, 1885.

Application filed January 9, 1880.

To all whom it may concern :

Be it known that we, WILLIAM E. SAWYER, a resident of the city, county, and State of New York, and ALBON MAN, a resident of Brooklyn, in the county of Kings and State aforesaid, both citizens of the United States, Improvements in Electric Lamps, of which jointly have invented certain new and useful improvements, the following is a specification.

Our invention, speaking generally, relates to that class of electric lamps employing an incandescent conductor inclosed in a transparent, hermetically-sealed vessel or chamber, from which oxygen is excluded, and constitutes an improvement upon the apparatus shown in Letters Patent No. 205,144, granted to us June 18, 1878.

Our invention relates more especially to the incandescing conductor, its substance, its form, and its combination with the other elements composing the lamp. Its object is to secure a cheap and effective apparatus; and our improvement consists. first, of the combination, in a lamp-chamber composed wholly of glass, as described in Patent No. 205,144, of an incandescing conductor of carbon made from a vegetable fibrous material, in contradistinction to a similar conductor made from mineral or gas carbon, and also in the form of such conductor so made from such vegetable carbon, and combined in the lighting circuit within the exhausted chamber of the lamp.

The accompanying drawings show all our improvements embodied in an apparatus or lamp substantially like that represented in the patent above referred to, being the form in which we have practically used it; but some of our improvements may be used in connection with other forms of lamps with equally good effect.

Fig. 1.

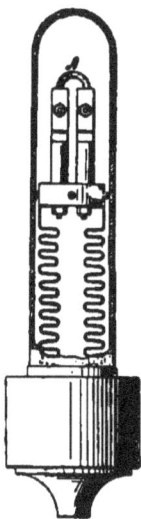

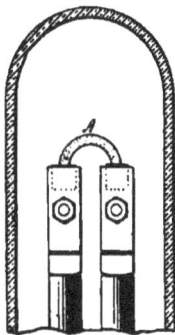

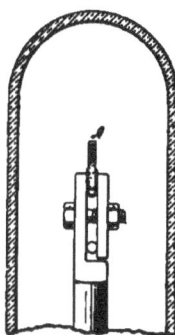

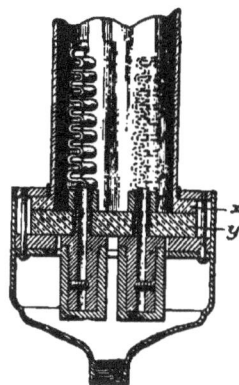

Fig. 3.   *Fig. 4.*

Fig. 2.   *Fig. 5.*

Reference being had to said drawings, Fig. 1 is a top view of the lamp ; Fig. 2, a side elevation thereof; Fig. 3, a side view in elevation of the burner on an enlarged scale, to show its details more clearly, and Fig. 4 is a similar edge view.

Fig. 5 of the drawing shows a vertical section through the bottom of the lamp. In this figure *x* is a glass flange on the bottom of the lamp-chamber, *y* is a glass disk corresponding in size to the flange, and is ground to the bottom thereof to form an air-tight joint, so that the entire wall of the chamber is formed of glass, the electrodes passing through the glass disk in the manner shown to form the lighting-circuit in the chamber, substantially as in said Patent No. 205,144. The scaling of the electrodes, where they pass through the glass wall, is done with any suitable cement, or in any of the well-known methods of sealing glass upon metal electrodes previous to the filing of this application. [1]

The electric connections of this lamp are made in the base thereof, substantially the same as in our Patent No. 210,809, dated December 10, 1878, and the whole bottom is inclosed in a cup filled with wax or other suitable cement, the same as in that patent, the cement sealing in this lamp being also applied in substantially the same way as in the patent last above mentioned, the invention making the subject-matter of this application being improvements upon the lamps described in the patents above referred to, to the extent of the claims making part hereof.

In the practice of our invention we have made use of carbonized paper, and also wood carbon. We have also used such conductors or burners of various shapes, such as pieces with their lower ends secured to their respective supports and having their upper ends united so as to form an inverted V-shaped burner. We have also used conductors of varying contours—that is, with rectangular bends instead of curvilinear ones; but we prefer the arch shape.

No especial description of making the illuminating carbon conductors, described in this specification and making the subject-matter of this improvement, is thought necessary, as any of the ordinary methods of forming the material to be carbonized to the desired shape and size, and carbonizing it while confined in retorts in powdered carbon, substantially according to the methods in practice before the date of this improvement, may be adopted in the practice thereof by any one skilled in the arts appertaining to the making of carbons for electric lighting or for other use in the arts.

An important practical advantage which is secured by the arch form of incandescing carbon is that it permits the carbon to expand and contract under the varying temperatures to which it is subjected when the electric current is turned on or off without altering the position of its fixed terminals. Thus the necessity for a special mechanical device to compensate for the expansion and contraction which has heretofore been necessary is entirely dispensed with, and thus the lamp is materially simplified in its construction. Another advantage of the arch form is that the shadow cast by such burners is less than that produced by other forms of burners when fitted with the necessary devices to support them.

Another important advantage resulting from our construction of the lamp results from the fact that the wall forming the chamber of the lamp through which the electrodes pass to the interior of the lamp is made wholly of glass, by which all danger of oxidation, leakage, or short-circuiting is avoided.

The advantages resulting from the manufacture of the carbon from vegetable fibrous or textile material instead of mineral or gas carbon are many. Among them may be mentioned the convenience afforded for cutting and making the conductor in the desired form and size, the purity and equality of the carbon obtained, its susceptibility to tempering, both as to hardness and resistance, and its toughness and durability. We have used such burners in closed or hermetically-sealed transparent chambers, in a vacuum, in nitrogen gas, and in hydrogen gas; but we have obtained the best results in a vacuum, or an attenuated atmosphere of nitrogen gas, the great desideratum being to exclude oxygen or other gases capable of combining with carbon at high temperature from the incandescing-chamber, as is well understood.

---

[1] For an example of this method of sealing, see description and figure of Crookes' radiometer, pp. 27, 28 *ante.*

The nature of our inventions and the operation of our improved lamp will be readily understood from the foregoing description and the following claims. We claim as our joint invention—

1. An incandescing conductor for an electric lamp, of carbonized fibrous or textile material and of an arch or horseshoe shape, substantially as hereinbefore set forth.

2. The combination, substantially as hereinbefore set forth, of an electric circuit and an incandescing conductor of carbonized fibrous material, included in and forming part of said circuit, and a transparent harmetically-sealed chamber in which the conductor is inclosed.

3. The incandescing conductor for an electric lamp, formed of carbonized paper, substantially as described.

4. An incandescing electric lamp consisting of the following elements in combination: first, an illuminating-chamber made wholly of glass hermetically sealed, and out of which all carbon-consuming gas has been exhausted or driven; second, an electric-circuit conductor passing through the glass wall of said chamber and hermetically sealed therein, as described; third, an illuminating conductor in said circuit, and forming part thereof within said chamber, consisting of carbon made from a fibrous or textile material, having the form of an arch or loop, substantially as described, for the purpose specified.

In testimony whereof, we have hereunto subscribed our names this 8th day of January, 1880.

WILLIAM EDWARD SAWYER.
ALBON MAN.

After the Sawyer-Man patent had been issued, the attorneys of Edison, fully appreciating the controlling importance of the subject-matter of its claim, demanded from the Patent Office the institution of a new interference, with specific reference to the fibrous carbon claims of the Sawyer-Man patent, which Edison had now included in an amendment of his original specification; but such action was refused by the Office on the 27th of June, on the ground that the question had already been passed upon in the original interference.

This final decision terminated the long, tedious and stubbornly-contested conflict over the question of priority of invention of the arch-shaped fibrous carbon, which had occupied the attention of the Patent Office for a period of four years and nine months.

COMMERCIAL INTRODUCTION OF THE INCANDESCENT LAMP.

Having thus traced the course of the proceedings respecting the legal title to the most essential and characteristic feature of the modern incandescent lamp, it will be proper to go back to 1880, and trace in like manner, the process of its commercial development.

While the exhibition of Edison's lights was going on at Menlo Park, in January, 1880, in response to a series of questions addressed to him by a representative of one of the leading journals of New York city, Edison said that he considered his work on the electric light practically finished, although he was still experimenting with a view to reducing its cost. He claimed to get an amount of light equivalent to 10 ordinary gas-jets per horse-power. The engine used at Menlo Park was of eighty horse-power, and the cost of running it was 75 cents per hour. Twenty-three lamps had been burning continuously from Friday until Wednesday, and thirty-three from Wednesday until 10 o'clock on Thursday night. During this time the engine was stopped for an hour to take water. Not a lamp had been injured, and all were regulated at the central station. Three of these lights were situated at a distance of one-fifth of a mile.

The public exhibition at Menlo Park had already, in the minds of intelligent and competent critics, established beyond question the practicability of incandescent lighting. From this on it was merely a question of economy. In other words, it remained to be determined whether the light from incandescent lamps could be furnished to the public at a profit to the manufacturer, and at a price which the consumer could afford to, or would be willing to pay.

The solution of this problem depended mainly upon two separate factors; first, the original cost of each lamp plus the cost of renewals; and second, the cost of maintaining each lamp in its proper condition of incandescence. A very brief experience was sufficient to demonstrate that these two conditions were necessarily antagonistic to each other; that is to say, the highest economy in the cost of lamps and renewals could only be secured by the expenditure of a disproportionate amount of energy. On the other hand, high incandescence, involving a small expenditure of energy per unit of light, materially shortened the life of the lamp, and thus materially added to the frequency and cost of renewals.

One of the first, if not the first announcement, of the results of an investigation of these points, was in a lecture delivered before the Franklin Scientific Society in Philadelphia, on the evening of March 24th, 1880, by Professor G. F. Barker, who had at that time, by invitation of Edison, in conjunc-

tion with Professors Young, Brackett and Rowland, just completed a series of measurements of the efficiency of his lamp.
Professor Barker said :

The vacuum in the lamp is now within one four-millionths of a perfect
exhaustion. . . . . A long series of experiments with various kinds of
fibres, such as southern moss, basswood, palmetto, Mexican hemp, jute, cocoa-
nut, palm and manilla, demonstrates that manilla fibre is much tougher and
better adapted for horse-shoes in the lamp than any other material. . . . .

. . . I now propose to show that we get our light energy (for there is
nothing that is bought or sold in all this world that does not represent energy)
cheaper through electricity than gas. To calculate with exactness the comparative merits of electricity over gas, we must reduce it to heat. This is done by
burning one of these lamps immersed in water. This is called the test by the
calorimeter. . . . .

One pound of coal gas will give us light equal to 75 candles. This pound,
if burned in the calorimeter, that is, under water, will equal 20,000 heat units,
equal to 16,000,000 foot-pounds, equal to 250,000 foot-pounds per minute.
One candle, therefore, requires more than 3,300 foot-pounds per second.
Again, mechanically, the cost of running an engine may be very liberally set
down as three cents per horse-power per hour. This applied to driving Edison's
generator, gives us 10 lights of 16 candles each; that is, 10 lights per three
cents, or 3 of one cent for each light of 16 candle-power.

Now look at the contrast. Gas, we will say, is $2.00 per thousand. Five
cubic feet of gas costs at that rate, one cent. That is to say, we get three
electric lamps at the same cost as one five-foot gas-burner light. In language
still more simple, I can positively say, after the most careful computation, into
which I cannot go on this occasion, that until gas can be furnished for sixty
cents per one thousand cubic feet, the electric light is cheaper.[1]

The results of the experiments and measurements which
had been made by Professors Barker, Brackett, Young and
Rowland, were published in the form of a paper in the *American Journal of Science,* and also in advance in the *New York
Herald* of March 27, 1880. The results of the calorimeter and
photometer test with four different lamps, gave respectively as
the mean number of 16 candle-power lights per indicated
horse-power, 4.8, 8.9, 5.8, and 9.2. These figures are stated to
represent 70 per cent. of the computed value, 30 per cent.
being allowed for engine friction, loss of energy in the generator, and the heating of wires. The conclusion of the paper
is as follows :

The increased efficiency with rise of temperature is clearly shown by the
table, and there is no reason, **provided the carbons can be made to stand,** why the
number of candles per horse-power might not be greatly increased, seeing that

---

[1] *New York Herald,* March 25, 1880. It will be observed that the lecturer has fallen into
the error of comparing the *net cost* of production of the electric light with the *selling price* of
coal gas.

the amount which can be obtained from the arc is from 1,000 to 1,500 candles per horse-power. *Provided the lamp can be made either cheap enough or durable enough, there is no reasonable doubt of the practical success of the light;* but this point will evidently require much further experiment before the light can be pronounced practicable.

An independent investigation of the Edison lamp was undertaken shortly afterwards by Professors Henry Morton, Alfred Mayer and B. F. Thomas, of the Stevens Institute of Technology. The single lamp used was furnished by the proprietors of the *Scientific American,* and the results were published in that journal April 17, 1880. The following extracts contain the important parts of this paper :

The current from the battery was divided into two branches, which traversed, in opposite directions, the two equal coils of a differential galvonometer. One branch then traversed the lamp, while the other passed through a set of adjustable resistances composed of German-silver wires stretched in the free air of the laboratory, to avoid heating. (Careful tests of these resistances showed that no sensible heating occurred under these circumstances).

Matters being thus arranged, the resistances were adjusted until the galvanometer showed no deflection, when the candle-power of the lamp was taken repeatedly in the photometer, and the amount of resistance was noted.

These measurements were several times repeated, shifting the coils of the galvanometer and reversing the direction of the current.

The results so obtained were as follows:

Resistances.		Condition of Loop.
123 Ohms	....	Cold.
94 "	....	Orange light.
83.7 "	....	9-10 candle.
79.8 "	....	5 "
75 "	....	18 "

The photometric measurement was in all these cases taken with the carbon loop at right angles to the axis of the photometer, which was, of course, much in favor of the electric lamp. On turning the lamp round so as to bring the carbon loop with its plane parallel with the axis of the photometer disk, the light was greatly diminished, so that it was reduced to almost one-third of what it was with the loop sideways to the photometer disk.

Having thus determined the resistance of the lamp when in actual use, it was next desirable to measure the quantity of the current flowing under the same conditions.

To do this the current from fifty cells of battery was passed through a tangent galvanometer as a mere check or indicator of variations, and then through a copper voltameter, i. e., a jar containing solution of cupric sulphate with copper electrodes immersed, and then through the lamp placed in the photometer.

Under these conditions it was found that during an hour the light gradually varied from about 16 candles at the beginning to about 14 candles at the end, making an average of about 15 candles, measured with side of loop toward disk.

The galvanometer during this time only showed a fall of half a degree in the deflection of the needle.

Carefully drying and weighing the copper electrodes, it was found that the cathode had gained 1.0624 grammes.

Now, it is well known that a current of one weber [1] takes up 0.00326 gramme of copper per second, which would make 1.1736 grammes in an hour; therefore the current in the present case must have been on the average

$$\frac{1.0624}{1.1736} = 0.905 \text{ webers, or a little less than one weber.}$$

Having thus obtained the resistance of the lamp when emitting a light of 15 candles, namely, 76 ohms, and the amount of current passing under the same conditions, namely, 0.905 weber, we have all the experimental data required for the determination of the energy transformed or expended in the lamp, expressed in foot-pounds. For this we multiply together the square of the current, the resistance, the constant 0.737335 (which expresses the fraction of a foot-pound involved in a current of one weber traversing a resistance of one ohm for one second), and the number of seconds in a minute. Thus, in the present case we have $0.905^2 = 0.8125$, and $0.8125 \times 76 \times .737335 \times 60 = 2753.76$ foot-pounds.

Dividing these foot-pounds per minute by the number of foot-pounds per minute in a horse-power, that is, 33,000, we have 0.08, that is, about eight one-hundredths or one-twelfth of a horse-power, as the energy expended in each lamp.

It would thus appear that with such lamps as this, one-horse power of energy in the current would operate twelve lamps of the same resistance with an average candle-power of 10 candles each, or 120 candles in the aggregate. [2]

Assuming that Siemens or Brush machine were employed to generate the electric current, such current would be obtained, as has been shown by numerous experiments, with a loss of about 40 per cent. of the mechanical energy applied to the driving-pulley of the machine. To operate these 12 lamps, therefore, we should have to apply more than one horse-power to the pulley of the machine, so that when this loss in transformation had been encountered there should be one horse-power of electric energy produced. This would call for 1 2-3 horse-power applied to the pulley of the dynamo-electric machine, by the steam engine.

---

[1] The "weber" referred to is the unit of electric current now known as the "ampere."

[2] The average of the maximum and minimum lights in azimuths at right angles and in the plane of the loop was taken as the average luminous power of the lamp. Our reason for this, was, that we found by measuring the light at every azimuth varying by ten degrees between 0° and 180°, this was approximately the true expression for the total amount of light emitted. We see from the article of Professors Rowland and Barker, in the *American Journal of Science*, that they, assuming certain conditions and discussing the same in a mathematical manner, have reached a different result; but as experiment shows this result not to be attained in fact, it is evident that the assumptions on which the mathematical reasoning is based do not include all the conditions present in the experiment.

Two other sets of experiments, made since those given in our paper of April 17, in which the candle power of the loop was in its best condition, 17.6 and 19.8 candles, corresponding to averages of 11.7 and 13.2 candles respectively, showed a consumption of energy of 0.104 and 0.109 horse-power per lamp, or 9.6 and 9.1 lamps per horse-power. This would give 112 candles and 120 candles respectively per horse-power of electric energy consumed or transformed in the lamp. These results certainly agree very closely with each other and with our former determinations. [Additional note in *Sci. Am.* (n. s.) xlii, 273.

To produce one horse-power in a steam engine of the best construction about 3 lbs. of coal per hour must be burned, and therefore for 1 2-3 horse-power 5 lbs. of coal must be burned.

On the other hand one pound of gas-coal will produce 5 cubic feet of gas, and will leave, besides, a large part of its weight in coke, to say nothing of other "residuals," which will represent practically about the difference in value between "steam making" and "gas making" coal, so that it will not be unfair to take 5 lbs. of gas coal as the equivalent of 5 lbs. of steam coal.

These 5 lbs. of gas coal will then yield 25 cubic feet of gas, which, if burned in five gas burners of the best construction, will give from 20 to 22 candles each, or 100 to 110 candles in the aggregate.

We have, then, the twelve Edison lamps producing 120 candles and the five gas burners producing 100 to 110 candles, with an equivalent expenditure of fuel·

If each apparatus and system could be worked with equal facility and economy, this would of course show *something* in favor of the electric light; but when in fact everything in this regard is against the electric light, which demands vastly more machinery, and that of a more delicate kind, requires more skillful management, shows more liability to disarrangement and waste, and presents an utter lack of the storage capacity which secures such a vast efficiency, convenience, and economy in gas, then we see that this relatively trifling economy disappears or ceases to have any controlling importance in the practical relations of the subject. [1]

It will be seen that the results of the measurements of the energy consumed in the lamp did not differ widely from some of those obtained by Professor Barker and his confreres. The final conclusion, however, is subject to an important correction. The efficiency of the dynamo machine is estimated at 60 per cent., but it afterwards appeared that the Edison machine, presumably the same one employed in Professor Barker's measurements, gave in fact an electrical efficiency of 90.7 per cent., and a net efficiency of 83.9 per cent., 24 per cent. above the figures assumed by Professor Morton. [2]

In the New York *Sun* of June 14, 1880, appeared an article, evidently prepared with much care, in which the writer undertook to state with as much accuracy as possible what had been accomplished by Edison in his efforts to solve the problem of electric lighting by incandescence, a question which up to that time had never been satisfactorily answered. The writer of the article says :

It has been difficult, amid the fierce disputes of the electricians over Mr. Edison's work, for an unscientific and unprejudiced on-looker to clearly see just what the inventor has accomplished. It is quite possible, however, by a careful study of the subject, to ascertain the real state of affairs. . . . .

[1] *Scientific American*, xlii, (n. s.) 241,273.

[2] Vide report of measurements of efficiency of Edison's dynamo, by Professors C. F. Brackett and C. A. Young. Apr. 10, 1880. *Sci. Am.* (n. s.) xlii, 308.

There are two questions, the answer to which will place the public in possession of all the facts that it needs or cares to know about the electric light. *First*—Has Mr. Edison invented a lamp by which electricity can be used satisfactorily in the place of gas and kerosene for lighting purposes ? And *second*, can he supply electricity for such purposes to the public cheaper than gas is supplied ?

1. The first question may be answered in the affirmative. There seems to be no doubt that Mr. Edison's electric lamp answers its purpose so far as it is a means of supplying light from electricity.

Whether it answers the purpose in the best way, or whether it does not need to be improved before it will be practically satisfactory, are subsequent questions not so easily answered. In fact, time alone can definitely answer them. Nevertheless, Mr. Edison's lamp does give light from electricity, and the light in itself is better than gaslight. . . . . . There is one fact about Mr. Edison's lamps that is disappointing to the general observer. They are not indestructible. All the lights that were started in Menlo Park last winter, with a sort of challenge to the world to come and behold the great problem solved, have burned out. The houses of Menlo Park that a few months ago rejoiced in the glow of electric illumination, are now dimly lighted by old-fashioned kerosene lamps, and the walks at night know no other light but that of the moon and the stars. The impression got abroad last winter that the inventor regarded the carbon horseshoes in the lamps as practically indestructible, and that as soon as he had invented some way to obviate the difficulty of the cracking of the globes from the heat, where the fine wires penetrate into the airless interior, all trouble about the durability of the lamps would be passed. Now Mr. Edison says he has gotten over the trouble about the cracking of the globes, and yet he is experimenting with other substances besides the paper previously used to make the carbon horseshoes. Mr. Edison explains that the public misunderstood the matter, and that he was not disappointed because the horseshoes burned up. He thought in advance that they would last on an average about 600 hours. He never supposed they would be indestructible. They really did last on the average 792 hours. These, he says, were the first ever made, and since then he has improved the processes of manufacture, so that when he starts new ones he is confident that their average life will be increased. Besides, he has found that carbon loops made of wood last twice as long as those made of paper. So he will probably substitute wood instead of paper for making the carbon horseshoes. The wood carbon lasts longer than the paper carbon because it is more uniform in texture and has fewer imperfections. Besides wood handles to better advantage, and is cheaper. Mr. Edison does not admit that his experiments with other substances besides paper for forming the carbon loops imply that, in that respect, his lamp is imperfect. Still the impartial observer can hardly avoid the impression that there is a loose screw here, and that great importance is attached to this point is shown by the tireless energy with which the search for the best substance to form the horseshoe has been pushed. In this search, first and last, he has fairly exhausted the resources of the famous gardens of Kew. He has made himself thoroughly acquainted with the properties of a vast number of vegetable substances. . . . He says the object is merely to get the very best material for the purpose before the lamps are introduced in practice. So, according to Mr. Edison, the object is to improve to the last degree before the light is given to the public. But the impression that the outsider gets is that the effort is not so much to round off a finished work as to complete a work that is not yet satisfactorily finished.

The question suggests itself, if the lamps are limited in duration, can they be easily and cheaply replaced, so that their wearing out will be neither an annoyance nor a burden of expense to the consumer ? Mr. Edison says yes. With his new carbon and improved processes of manufacture he calculates to double the life of the lamps. Estimating the average time during which lights are required, he calculates that if his lamps will burn uninterruptedly for 1,600 hours they will last for family use about a year. Those that were tried in Menlo Park last winter averaged about 790 hours, and according to his calculations would have lasted a family about six months. They can be replaced, he says, as easily as the simplest form of lamp chimneys. If one of his lamps becomes exhausted any servant can lift it off and put on a fresh one in a few seconds. According to the calculations of some of Mr. Edison's assistants, the lamps will be sold for 25 cents apiece at most, when the processes of manufacture are perfected, although they now cost a dollar to make. At that rate, if they last only six months, the cost of the lamps for a year would be only 50 cents. Mr. Edison calculates that they will last much longer than lamp chimneys do. An ordinary lamp chimney, however, may last a man's lifetime, or it may be broken within an hour; that is a question of comparative care, and the destruction, when it comes, comes from without. But, it seems, the electric lamps are limited in duration by a law of their being, so to speak ; their destruction comes from within.   .   .   .   .

2. The second question, Can Mr. Edison supply electricity for lighting purposes to the public cheaper than gas is supplied is more difficult to answer. Mr. Edison says he certainly can, and he adds that he is disposed to tell the public just what his light costs, but the capitalists who are backing him say no ; and so all that he is permitted to say is, that no matter what the price of gas will be, the electric light will always be supplied a little cheaper. What he is doing now, he says, is not for the public but for about a dozen men who compose the electric light company. He admits that the display at Menlo Park, that is to be made in the summer, is to be for the sake of demonstrating to capitalists what can be done, but he avers that does not admit the inference that they are dissatisfied with the results already shown. He claims that the reports of the scientific men who personally examined his invention, to the effect that from his generating machine 83 per cent, available efficiency is obtained, and that from every horse-power he gets from 7 to 13½ lights, according to their intensity, taken in connection with the average cost per horse-power, demonstrates that his light can be supplied cheaper than gas.[1] In regard to the electric stock, Mr. Edison says of the 3,000 shares only 40 were ever speculated with, and that only by the small-fry holders, who were excited by the experiments of last Winter. He yet has every share he ever held, and so have the other heavy stockholders.   .   .   .   .

The answer to the second question, then, seems to be that according to Mr. Edison's calculations, based on the results achieved on a comparatively small scale in Menlo Park, the electric light can be furnished to the public cheaper than gaslight; but the capitalists who have invested their money in the company are not so well assured that this would prove true in practice on a large

---

[1] Vide report of Professors Barker, Brackett and Young, p. 65, *ante.*

scale, that they do not require further practical demonstrations of its truth, such as the display to be made in the summer is designed to supply. Mr. Edison has yet to demonstrate to them that with a lamp, confessedly destructible even without accident, he can keep New York city lighted better and cheaper than the gas companies can. So long as those who are peculiarly interested are not convinced to the bottom of their pockets, the public will have to wait.

The steamer *Columbia*, that sailed from this city a few weeks ago for California, was completely lighted by Mr. Edison's lamps. A lamp of an intensity of about five-candle power was used for the state-rooms. The *Columbia* is to run between San Francisco and the Columbia River.

The first commercial electric lighting plant which was installed by the Edison company and put in operation in the hands of outside parties was the one above referred to, on board the steamship *Columbia*, built in Chester, Pa., in 1879, for the Oregon Railway and Navigation Company.

The history of this plant is briefly given in the following letter from the engineer of the company, under date of February 24, 1882 :

In 1879, while the *Columbia*, which contains a large number of passenger rooms, was under construction, President Villard conceived the idea of lighting each room in the vessel independently by the electric light. Thereupon, at your suggestion and by his orders, I wired the ship with No. 11 wire for mains and No. 32 wire for loops, insulated by double cotton paraffine and painted over all. The weres were run throughout the entire vessel, but the project at that time being experimental, we lighted only the passenger rooms and main saloons. The dynamos, of which we had four, one of them running at half of the speed of the others as an exciter or fielder, were of the class you now call "A," and were all run from a countershaft directly overhead, driven in turn by a pair of vertical engines at a very high angle in order to economize freight space. On the night of the 2d of May, 1880, we started up the dynamos, and from the time when the steam was first turned on until the present day they have worked to our entire satisfaction under all circumstances.

We found the light of the greatest value for the examination of the ship's propeller, rudder or hull, which examination we conducted by connecting to a main line aft, or at any convenient point, a coil of insulated wire with lamps attached to a sinker.

The first lamps used, being of the paper carbon variety, were irregular in their duration of life and so liable to breakage by heavy shocks that I found it best to suspend them from the wires above, and to do away with the sockets entirely. The lamps being surrounded with a ground globe, the attachment was hid, the lights being suspended from the ceiling. Since the arrival of the ship on the Pacific coast we have received a full supply of new bamboo-carbon lamps. How well these have worked can best be seen from the following report of Chief Engineer Van Duzer : "I have now one hundred and fifteen lamps in circuit, and have, up to date, run four hundred and fifteen hours and forty-five minutes without one lamp giving out." [*]

---

[*] *Fourth Bulletin of Edison Electric Light Co.*, Feb. 24, 1882, p. 1.

A local corporation known as the Edison Electric Illuminating Company was formed in 1880, with the object of introducing Edison's system of electric lighting on a large scale, in New York city. This company applied to the municipal government in December, 1880, for a franchise for opening the streets, which having been granted, operations were shortly afterwards commenced in the territory bounded by Wall, Spruce and Nassau streets and the East river. The conductors were embedded in insulating material, enclosed in wrought iron pipes laid eighteen inches below the surface.

While the work of preparation for this system was going on, which necessarily occupied considerable time, the Edison Company for Isolated Lighting, operating under the auspices of the Edison Electric Light Company, was formed early in 1882, and engaged in the installation of plants in isolated buildings, vessels and the like. Up to June, 1882, this company reported installations of such plants, aggregating 10,424 lamps.

The installation of the central-station plant in New York city was completed in the autumn of 1882, the machinery having been started for the first time on the afternoon of September 4, supplying 85 buildings wired for 2,323 lamps. The price charged consumers for the light was computed on the basis of one dollar per thousand candles, which would be equivalent to about $2.25 per thousand feet of gas.

In April, 1884, it was officially stated that the New York central-station plant was lighting 500 houses, and that 11,272 lamps had been connected with the mains. [1] The number of isolated Edison plants had, at the same date, increased to 307, with an aggregate of 59,173 lamps. [2]

The commercial manufacture of Edison lamps was commenced in November, 1880, at which time a lamp factory was started at Menlo Park as a separate department of the business. The works were removed to East Newark, N. J., April 1st, 1882, where the business has been continued since that time.

A local organization was formed in New York on July 16, 1883, under the name of the Sawyer-Man Illuminating Company, with the object of installing incandescent electric light-

---

[1] *Twenty-second Bulletin of Edison Electric Light Co.* p 2.   [2] Ibid. p. 5.

ing plants under the Sawyer-Man patents. After the decision of the Commissioner of Patents had been rendered affirming the priority of Sawyer and Man over Edison in the invention of the arch-shaped fibrous carbon, measures were taken to commence the commercial installation of plants, the first one being placed in the United Bank building in New York city in October, 1883. The type of lamp adopted by this company, although in outward appearace much resembling that of Edison, was like that lamp essentially nothing more than an adaptation of the fibrous carbon to the exhausted glass bulb of Crookes' radiometer. This will be apparent by com-

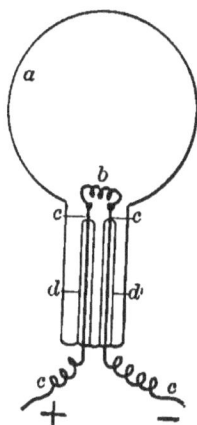

FIG. 1.
CROOKES' RADIOMETER.

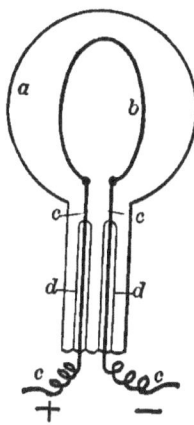

FIG. 2.
MODERN INCANDESCENT LAMP.

In both figures, *a* is the exhausted glass chamber, *b* is the illuminant or incandescing conductor, *c c* are the leading-in wires of platinum, to which the ends of the illuminant are attached, and which are hermetically sealed by fusion into glass tubes *d d*. The main conducting wires are attached at $+$ and $-$.

paring the two figures in the cut, figure 1 being a facsimile of a portion of the cut showing Crookes' apparatus, taken from one of the papers published by him in the *Philosophical Transactions* of the Royal Society, and figure 2 the same with an illuminant of fibrous carbon in an arch form, employed by the Sawyer-Man Company as a substitute for the platinum conductor of Crookes.

In the Edison lamp, the carbon illuminant has always been made comparatively long and thin, while the illuminants in all the early lamps of Sawyer and Man were comparatively short and thick. It appears from evidence given in the interference proceedings that in their individual ideas, Sawyer and

Man were at variance with each other in respect to the best form and proportion of the carbon, Sawyer favoring the short thick carbon of low resistance, and Man the longer and thinner carbon of higher resistance.[1] This long and thin, or "filamentary" carbon, presenting, as a necessary consequence, a high resistance to the electric current, has always been claimed as one of Edison's most important inventions.[2]

While there can be but one opinion as to the great practical merit of this conception, it would nevertheless appear to be one involving mechanical or engineering skill rather than invention in the proper sense of that term. An important fact in this connection must not be overlooked, and that is that it is impracticable to make a long and thin, or in other words, a "filamentary conductor," except of the carbonized fibrous material of Sawyer and Man, and commercially speaking, it is equally essential that it be disposed in an arch or horseshoe form. A patent was, however, taken out by Edison on January 27, 1880, evidently intended to specifically cover this feature of construction, the claims of which were as follows :

An electric lamp for giving light by incandescence, consisting of a filament of carbon of high resistance made as decribed and secured to metallic wires, as set forth.

The combination of carbon filaments with a receiver made entirely of glass, and conductors passing through the glass, and from which receiver the air is exhausted, for the purpose set forth.

The validity of these important claims has, however, never been passed upon by our courts, and owing to a peculiar chain of circumstances, it can now scarcely be expected that this will be done hereafter. The circumstances referred to are as follows :

The application for a United States patent was made by Edison on November 4, 1879, and at about the same time he

---

[1] In his testimony in the interference proceedings in *Sawyer and Man vs. Edison*, Sawyer testified, in his answer to cross-question 42, as follows :

Mr. Man's idea and mine upon that question of resistance began to differ ; that is to say, my theory that the most perfect electric lamp would be one in which the incandescent conductor had the highest resistance and the least transverse mass was gradually abandoned by myself, while largely retained by Mr. Man. My present theory being that the most perfect electric lamp is one in which the incandescent conductor has not only the least transverse mass, but the least resistance.

[2] In a card published by Edward H. Johnson, President of the Electric Light Company, May 24, 1885, he says :

Edison's patent of January 27, 1880, [see above] applied for at the time of his great discovery, covers broadly all the elements of that discovery, and is, therefore, fundamental and controlling.

made application for a patent on the same invention in the Dominion of Canada. The last named patent was granted on November 17, 1879, while the United States patent did not issue until January 27, 1880. The patent laws of the United States provide that when patents for an invention have been obtained in any foreign country prior to the domestic patent, that the latter shall expire at the same time as the foreign patent having the shortest term.[1] The consequence of this has been that the suits brought by the Edison company under this patent against the Sawyer-Man and other companies making or using incandescent lamps claimed to be of "high resistance," have been met by the plea that Edison's patent had expired by virtue of the provisions of the Federal statute. The language of the section is apparently clear enough, but it has nevertheless proved to be susceptible of varying interpretations, leading to widely different results. But in January, 1889, the Supreme Court of the United States rendered a decision in the appealed case of the *Bate Refrigerator Company* vs. *Hammond*, in which the following interpretation was put upon it:

These provisions of the Act of 1870, and the Revised Statutes, mean that the United States patent shall not expire so long as the foreign patent continues to exist, not extending beyond seventeen years from the date of the United States patent, but shall continue in force, though not longer than seventeen years from its date, so long as the foreign patent continues to exist. . .

. . . It is to be in force as long as the foreign patent is in force. . . . A contrary view to this has been expressed by several Circuit Courts of the United States. . . . The time of the expiration of the foreign patent may be shown by evidence *in pais*, whether the record of the foreign patent itself, showing its duration, or other proper evidence.[2]

It was claimed on behalf of the Edison Company that the validity of certain Edison patents, as to which a question had arisen whether they had or had not expired under the statute, was affirmed by this decision,[3] but subsequent events

---

[1] Section 4887 of the Revised Statutes of the United States says :—

Every patent granted for an invention which has been previously patented in a foreign country shall be so limited as to expire at the same time with the foreign patent, or, if there be more than one, at the same time with the one having the shortest term, and in no case shall it be in force for more than seventeen years.

[2] *U. S. Reports,* cxxi., 151.

[3] In an advertisement published in the New York *Electrical World* of February 2, 1889, signed by Edward H. Johnson, President of the Edison Electric Light Company, the following statement was made:

The Edison Electric Light Company takes pleasure in announcing to its patrons and the general public, that the decision of the Supreme Court of the United States, filed on the 21st inst., in the case of the Bate Refrigerator Company, is in fact a decision sustaining the Edison patents.

showed its effect to be quite different from what had been anticipated.

This decision was followed within a few weeks by a decision of the Deputy Commissioner of Patents of Canada in the case of the *Royal Electric Company* vs. *The Edison Electric Light Company*,[1] declaring Edison's patent of November 17, 1879, null and void for non-compliance with certain statutes of Canada in respect to manufacture and importation.[2] This was followed by a decision of Judge Thayer at St. Louis, May 25, 1889, in the case of *Huber vs. Nelson Manufacturing Company*,[3] in which it was held that the life of the domestic patent is measured by the actual duration of the foreign patent, and may be abridged as well as lengthened by circumstances which operate under the foreign law to abridge or lengthen the foreign monopoly. On August 8, 1889, the case of *Pohl vs. The Anchor Brewing Company* was decided by Judge Wallace, which defined the application of the doctrine enunciated by the Supreme Court in *Bate Refrigerator Company* vs. *Hammond*, in a case in which the circumstances were precisely the same as those of the Edison filament patent. Judge Wallace in his opinion said :

> The statute is capable of the meaning that the exclusive right to the invention here is to cease with the exclusive right of the patentee in any foreign country, or of the meaning that it shall continue to exist for such period, not exceeding seventeen years, as coincides with the shortest term of any foreign patent. The Supreme Court seems to have adopted the first of these meanings. This is the view expressed in the recent case of Huber vs. N. O. Nelson Manufacturing Company.

It would seem therefore that there is little probability that Edison's patent for a carbon filament of high resistance can ever come before our courts in such shape as to be considered on its merits. On many accounts this is to be regretted ; it is

---

[1] A full abstract of this decision, which was filed February 26, 1889, may be found in the *Electrical Engineer*, April 1889, vol. viii., p. 199.

[2] The Revised Statutes of Canada [Cap. 61, S c. 87] contain the following provision :

Every patent granted under this act, shall be subject and be expressed to be subject to the condition that such patent and all the rights and privileges thereby granted shall cease and determine, and that the patent shall be null and void at the end of two years from the date thereof, unless the patentee or his legal representatives, within that period, commence, and, after such commencement, continuously carry on in Canada, the construction or manufacture of the invention patented, in such manner that any person desiring to use it may obtain it or cause it to be made for him, at a reasonable price, at some manufactory or establishment for making or constructing it in Canada—and that such patent shall be void if, after the expiration of twelve months from the granting thereof, the patentee or his legal representatives or his assignee for the whole or a part of his interest in the patent imports or causes to be imported into Canada, the invention for which the patent is granted ; and if any dispute arises as to whether a patent has or has not become null and void under the provisions of this section, such dispute shall be decided by the Minister or the Deputy of the Minister of Agriculture, whose decision in the matter shall be final.

[3] *Federal Reporter*, xxxviii, 830.

particularly desirable that the line of demarcation between those improvements which unquestionably involve invention, and those which really exhibit nothing beyond an unusually high order of mechanical or engineering skill, should be more distinctly defined. The question is at best, a difficult one ; perhaps in its application to individual cases the most difficult one which the courts sitting in patent cases are ever called upon to determine.

The United States Electric Lighting Company, from the first, never sought to identify itself with the work of any particular inventor to the exclusion of others, but on the contrary adopted the policy, the wisdom of which ultimately became apparent, of acquiring the control of inventions or processes by whomsoever invented, or wheresoever found, which appeared to be valuable, or likely to become so in the future. In pursuance of this policy, it acquired at an early date inventions and patents, not only of Mr. Maxim, but of Professor Moses G. Farmer, Edward Weston and many other early and more or less successful workers in the field of incandescent lighting. A year or two after the period which has been spoken of, Mr. Maxim discontinued his electrical researches and henceforth devoted himself to the perfecting of his since famous machine gun, but not before he had devised and worked out a number of inventions and processes, mainly connected with hydro-carbon treatment, which have since proved to be of great importance and value. The United States Company, after an existence of some ten years, during which period it succeeded in building up an extensive and valuable business in electric lighting throughout the country, in May, 1889, passed by lease under the control of the Westinghouse Electric Company of Pittsburg, which had previously in a similar manner become possessed of the patents and inventions of Sawyer and Man.

HYDRO-CARBON TREATMENT OF INCANDESCENT CARBONS.

Reference has been made to the discovery, made by Sawyer and Man in 1878, that carbon conductors for use in incandescent lamps were capable of being greatly improved by electrically heating them in the presence of hydro-carbon, and thus causing carbon to be deposited upon them.[1] It is described in the patent of Sawyer and Man No. 211,262, of Jan-

---

[1] Vide p. 32, *ante*, text and note.

nary 7, 1879. The importance of the discovery was not so well appreciated at the time as it was after the incandescent lamp had begun to come into commercial use, and its conditions of economy and efficiency had been more carefully observed and studied. It was soon discovered that one of the principal factors affecting the economy of the incandescent lamp is the character of the radiating surface of the illuminant, and that perhaps the most valuable and important result of the hydro-carbon treatment consisted in its capacity for giving to that surface the highest possible power of luminous radiation.

The advantages secured by hydro-carbon treatment have been well stated by Major General C. E. Webber, as follows :[1]

In a good glow lamp, inseparable from uniformity of resistance are uniformity of surface and incandescence. Uniformity of resistance when hot, means that each lamp will (so to speak) consume the same amount of current, and this can only be secured by electrical pressure. Uniformity of incandescence and surface means that lamps possessing those qualities give almost the same candle power, and this can only be obtained in the manufacturing processes by visual measurement.

To secure all three, it is necessary to combine a means of visual and electrical measurement. The former is but approximate, the latter is as accurate as a Thomson's galvanometer can make it. In either case, however, when large numbers of filaments have to be standardized, neither measurement can be expected to produce greater accuracy than is essentially necessary for a good commercial lamp; and the fact is that while uniformity of E. M. F. is the condition which must not be departed from, no harm arises in practice if the current consumed and the candle power are not absolutely uniform. Thus the required resistance of each lamp is such a resistance as will, under the condition of uniform E. M. F. or electrical pressure, give, with a small varying area of surface, such a candle power as that, to the eye, all the lamps are alike when they are incandesced.

This required resistance is obtained by a process known in the electrical world as "flashing," a title which, I believe I am right in stating, owes its origin to Mr. Lane-Fox.

A patent for the process of hydro-carbon treatment (corresponding to their United States patent No. 211,262,) was obtained by Sawyer and Man in Great Britain, in the name of their agent, Frederick John Cheesbrough, (No. 4,847, of 1878), which ultimately passed by purchase into the hands of the Edison and Swan United Electric Light Company, Limited.[2] This company proceeded to bring an action for infringement

---

[1] Paper read before the Society of Arts December 8, 1886 ; *Tel. Journal and Elect. Review,* xix., 593.

[2] Vide statement of Arthur Shippey, in *Tel. Journal and Elect. Rev.,* xix., 204.

of the patent against Woodhouse and Rawson, prominent manufacturing electricians of London. The case was tried in May, 1886, before Mr. Justice Butt, in the High Court of Chancery. The defendants admitted the infringement, provided the patent was held to be valid, but contended that the process described by the patentee was not new, and had been fully anticipated by prior publications.

Justice Butt decided the case in favor of the plaintiffs, affirming the validity of the patent. In his opinion he said :

. . . . The patent is one for making the carbon conductors of electric lamps. **The process by which the desired result is obtained is, to my mind, one of singular beauty and efficacy.** The principal features of that process are these: a pencil of ordinary carbon is immersed in a hydro-carbon gas or liquid, a strong electric current is passed through the carbon, and in the course of its treatment the carbon is subjected to a gradually increased current of electricity. The defects in ordinary carbon for electric lighting, and the remedies applied by his process, are stated by the patentee in his specification. . . . *The peculiar merit of the invention consists in the well-nigh perfect adaptation of the means employed to the end aimed at, and this by the automatic action of the materials employed when subjected to intense heat generated by the passage of the electric current.* As pointed out, in an ordinary piece of carbon subjected to a strong electric current, there are observed points and lines of unequal brilliancy. The most brilliant parts are those where the carbon has the smallest sectional area, the resistance of the carbon to the passage of the electric current, and therefore the heat, being in inverse proportion to the size of the cross-section of the carbon. At these points of greatest heat, the deposition of the carbonaceous matter by means of the hydro-carbon gas is most active, and so equalizes the carbon, which is also made more dense by the operation.

After discussing at length the bearing upon the invention, of the alleged anticipating descriptions, he fails to find any description of the patentee's process, and concludes by saying :

I am, therefore, of opinion that the invention which Mr. Cheesbrough patented had not been anticipated by any prior publication of this process, that the patent sued on is valid, and, as its user by the defendant is admitted, there is infringement, for which the plaintiffs are entitled to relief.[1]

The development of the processes and methods of hydrocarbon treatment was taken up in this country at an early date by Hiram S. Maxim, under the auspices of the United States Electric Lighting Company. The first commercial incandescent lighting plant in New York City was started by this company in the autumn of 1880, in the reading rooms of the Safe Deposit Company, in the basement of the Equitable Life Insurance Building, at 120 Broadway. The plant consisted

---

[1] *Tel. Jour. and Elec. Review*, xviii, 501.

of about fifty 20-candle lamps, the illuminants of which had been subjected, in the process of manufacture, to hydro-carbon treatment. The following notice of this plant appeared in the New York *Evening Post* of November 12, 1880 :

In the reading rooms of the Safe Deposit Company, in the basement of the Equitable Life Insurance Building, at 120 Broadway, may now be seen the first successful attempt to use the electric light from incandescent carbon. Edison made last winter, as every one will remember, a display of lamps which resembled those now shown in the Equitable Building, but none of the lamps at Menlo Park gave more than a light equal to ten candles. The other defects of the Edison lights were that the power required was enormous, and that the carbon gave out sooner or later—often within a few days. One needs only to examine the lamps at No. 120 Broadway to see that something like perfection has been reached. In the reading room there are ten lights—a chandelier holding four lights and six single brackets. The light, in quality, is very much like that from a first-class oil lamp, steadier than gas, and of a yellow, clear, pleasant quality. There is an entire absence of the ghastly blue color and one of the flickering which makes the arc lights (to be seen in the basement corridor of the Equitable Building) so painful to the eyes if borne for any length of time. The only criticism which may be made, and which could only be made by an expert accustomed to examining electric lights, is that the pulsations of the engine are seen reflected in the changing intensity of the light, which falls and rises almost imperceptibly with every stroke of the engine which furnishes the power. In the vaults of the Safe Deposit Company there are fifty of these lamps, lighting up the whole place like daylight, night and day. Some of the lamps have been running for weeks ; others for a few days only. It is only a matter of time for the whole Equitable Building to be supplied in the same manner. The advantages of the light in the vaults, as one of the guardians explained, are that all possible danger of fire is avoided, and that the atmosphere, which, with the very low ceilings, becomes rapidly tainted by burning gas, remains perfectly cool and sweet.

The superior illuminating qualities of the Maxim lamps, as well as the fact that this was the first commercial installation which had been put in by an electric lighting company anywhere, with the exception of the plant furnished by the Edison company to the steamer *Columbia*, before noted, excited a marked degree of interest in scientific circles, which was not diminished by the publication within a few days of an interview with Professor G. F. Barker, of Philadelphia, in the *Evening Post* of November 22, 1880, in which he is reported as having said :

There is no doubt in my mind, or in that of Professors Morton and Draper, as to the value of Mr. Maxim's remarkable discovery. For years I have been an admirer of Edison's search for the true solution of the electric light problem, and I can testify to his unremitting energy and the exhaustive nature of his search. But another man found it.

I do not say that Maxim is a better electrician than Edison, but *he has invented a lamp which surpasses, I believe, even Edison's dreams.* Last week Professor Draper got several of Maxim's lamps and invited Professor Morton and me to go to his (Draper's) laboratory to test them. Edison had also been asked to send a few lamps, but wrote that an accident had happened to his machinery which would prevent him from doing so. He wrote, however, in answer to a question put by Professor Draper, that the highest light which he could get out of his improved lamps was fifty candle light. And from experiments which we have made with Edison's lamp we know that his lamps run at that intensity will not last for more than an hour.

We took these lamps of Maxim's and began our experiments. *As substitutes for gas burners of full power they were perfect.* We then began increasing the intensity until we ran one lamp for twenty-four hours at an intensity equal to that of six hundred and fifty candles, or the light given out by forty gas burners. The lamp was as good, apparently, when we finished as when we began. *This in my opinion, was the most remarkable performance of an incandescent lamp ever made.* Edison has a good generator, *but his lamp was old twenty years ago.* The hydro-carbon atmosphere of Maxim's lamp is new.

In 1885, Messrs. Siemens and Halske, manufacturing electricians of Berlin, Germany, made public the results of a series of tests which had been carried out by them for the purpose of determining the effect of the hydro-carbon treatment upon the efficiency and life of incandescent lamps. The comparison was made between a number of 16-candle lamps of their own manufacture, in which the carbon had been treated by the process heretofore described, and an equal number of Edison 16-candle lamps, containing illuminants which had received no treatment after having been removed from the carbonizing chamber. The two sets of lamps were burned under precisely the same conditions 9 hours per day, during an aggregate period of 800 hours, the potential at the lamp terminals being kept as nearly constant as possible.[1]

The results of the tests of Siemens and Halske show a marked difference between the two sets of lamps, especially with reference to efficiency during the earlier period of working. The efficiency was found to fall off very rapidly in the case of the untreated carbons, but remained comparatively constant with the treated carbons. Both sets of lamps were compared under electromotive forces higher and lower than the normal, but with substantially accordant results. When run at the normal potential of 100 volts, the initial efficiency of the treated and untreated carbons was respectively 333 and 259 candles per h. p. During the first 100 hours, the efficiency

---

[1] *Electrotechnische Zeitschrift*, November and December, 1885.

of the untreated carbons fell off very rapidly, becoming at the end of that time only 173 as against 300. At the end of 800 hours the figures were 210 and 148 candles per h. p. The following tables show the results in full:

Lamps with Untreated Carbons.

Burning Hours.	E. M. F. in Volts.	Current in Amperes.	Light in Candles.	Candles per electric H. P.	Watts per Candle.	Remarks.
0	100	.687	24.25	259	2.8	After 500 hours
100	100	.666	15.7	173	4.2	1 lamp burned
200	100	.666	15.3	169	5.0	thro.' The
300	100	.664	15.2	167	4.4	others sur-
400	100	.653	14.7	165	4.5	vived.
500	100	.640	13.7	157	4.7	
600	100	.634	13.3	154	4.8	
700	100	.630	13.1	153	4.9	
800	100	.620	12.5	148	5.0	

Lamps with Treated Carbons.

Burning Hours.	E. M. F. in Volts.	Current in Amperes.	Light in Candles.	Candles per electric H. P.	Watts per Candle.	Remarks.
0	100	.552	25.0	333	2.2	All 10 lamps
100	100	.550	22.5	300	2.4	survived.
200	100	.550	22.0	290	2.5	
300	100	.548	21.0	282	2.6	
400	100	.547	20.0	270	2.7	
500	100	.545	18.6	251	2.9	
600	100	.545	17.1	231	3.2	
700	100	.540	16.0	218	3.4	
800	100	.532	15.3	210	3.4	

The results given in the two tables may be summed up as follows:

	Carbons Treated.	Carbons Untreated.
Decrease in efficiency, (per cent,)	36.9	57.10
" " mean illuminating power (per cent,)	38.8	51.50
Mean illuminating power (c. p. per lamp)	19.67	14.91
Maximum difference in do. (c. p. per lamp)	9.7	11.75
Mean efficiency of lamps, (c. p. per electrical h. p.)	264.1	167.70
Increase in resistance (per cent,)	3.8	10.80

These results clearly establish the fact that over 30 per cent more light can be produced with a given expenditure of energy, with treated than with untreated carbons, and, moreover, that the latter maintain their efficiency much more uniformly during the ordinary lifetime of the lamp (800 hours).

THE COMMERCIAL MEASUREMENT OF ELECTRIC CURRENTS.

One of the most important points in connection with the distribution of electric energy for lighting and other purposes is that which has to do with the measurement of the energy consumed. Among the early inventions of Sawyer was one embracing this most important requisite of a general system of distribution. An application for a patent for this invention was filed October 17, and patented November 19, 1878. This appears to be the first apparatus of a practical character designed for the purpose. The following description and cut of the Sawyer and Man meter was published in the *Scientific American* of December 7, 1878.

Sawyer's Meter for Electric Lighting.

It is a simple clock arrangement, with an attachment designed to throw the dial hands into connection when a light is on. From each switch a pair of conducting wires are run to opposite studs on the wooden disk shown at the top of the figure. When no current passes through the lamp the revolving spring shown in front of the studded disk turns without making any record. When the current is on, the electric connection at each revolution is made through the pins assigned to the particular lamp, the armature of the magnet is moved, and the recording wheel is advanced one notch. This meter does not measure the quantity of electricity passing, but only the time a lamp is on. If two or any larger number of lamps are on, an equal number of connections are made at each revolution of the wheel and the record wheel is advanced to correspond. This registration is, of course, a mere matter of business detail.

As all modern incandescent lighting systems are based upon the principle of supplying to each lamp while in operation an electric current of prescribed potential and volume, it is evident that a registration of the time each lamp or group of lamps is burning affords an accurate measure of the consumption of energy, and therefore that this meter is a practical one; but long before the owners of the Sawyer and Man patents had occasion to use meters in central-station systems, not only had other and more economical and convenient apparatus been invented, but methods of distribution and consequently methods of measurement had been wholly changed by the introduction of the modern systems of distribution by alternating currents.

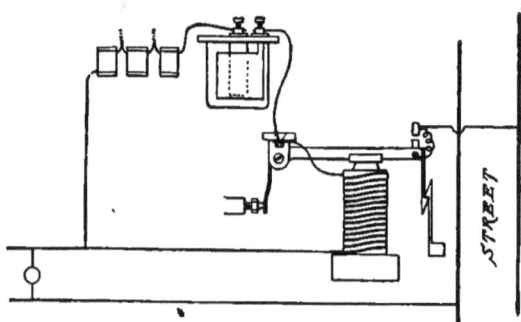

Edison's first Electrolytic Meter.

The necessity of some device for measuring the currents distributed was foreseen by Edison almost from the beginning of his experiments, but from his statement to a reporter published in the *New York Herald* of October 12, 1878,[1] it appeared that such a device was yet to be invented. Edison's first application for a patent on a device of this kind, was filed March 20, 1880, and patented December 27, 1881.

The principle of this meter is shown in the figure. It consists of an electrolytic cell placed in a shunt circuit, the resistances being so proportioned that a determinate fraction of the total current to be registered passes through the cell. This principle was embodied by Edison in many different forms, and was ultimately brought to such a state of perfection as apparently to answer every requirement of the service.

---

[1] Vide p. 17, *ante*.

CONCLUSION.

The growth of electrical enterprises in the United States is perhaps without a parallel in the history of industrial development. From a commercial point of view, the inception of incandescent lighting dates back no further than 1880. Statistics collected by the Secretary of the National Electric Light Association and presented at its meeting at Niagara Falls on August 6, 1889, show that there are now no less than 2,704,768 incandescent lamps in service in the United States, an increase of considerably over 200,000 within the preceding six months. To supply this number of incandescent lamps with electricity would require about 300,000 h. p., which is equal to nearly one-tenth of the aggregate steam and water-power used for all purposes of manufacture throughout the United States, according to the census of 1880. The aggregate capital now actually invested in electrical industries, principally electric lighting, railway and power distribution, is estimated by the same authority, as not less than $275,000,000.

More than 350,000 lamps are now in service in connection with the central station plants established under the auspices of the Westinghouse Electric Company of Pittsburgh, all of which have been installed within a period of less than three years.

THE END.

INDEX.

www.ingramcontent.com/pod-product-compliance
Lightning Source LLC
Chambersburg PA
CBHW021946190326
41519CB00009B/1159